The Golden Age of Drive-Thru IT

The Golden Age of
Drive-Thru IT

The True Potential of IT and
How IT Sells Itself Short

KEDAR SATHE

iUniverse, Inc.
Bloomington

THE GOLDEN AGE OF DRIVE-THRU IT
THE TRUE POTENTIAL OF IT AND HOW IT SELLS ITSELF SHORT

iUniverse books may be ordered through booksellers or by contacting:

iUniverse
1663 Liberty Drive
Bloomington, IN 47403
www.iuniverse.com
1-800-Authors (1-800-288-4677)

ISBN: 978-1-4759-8300-5 (sc)
ISBN: 978-1-4759-8299-2 (hc)
ISBN: 978-1-4759-8298-5 (e)

Library of Congress Control Number: 2013905692

Printed in the United States of America.

iUniverse rev. date: 3/29/2013

DEDICATION

*I dedicate this book to my parents, my wife, Aruna,
my son, Shaunak, and my family in India.*

TABLE OF CONTENTS

ACKNOWLEDGMENTS

This book and my life would not have been possible without help, unique contributions, advice, and lessons I learned from the many wonderful people I have met in my simple life.

A special thanks to the many wonderful teachers who have had a positive impact on me. Specifically, (late) Mr. Nagaraj, my 9th grade Computer Science teacher, and Dr. James D. Garber, my research advisor and head of Department of Chemical Engineering at University of Louisiana at Lafayette. Mr. Nagaraj and Dr. Garber truly imbued me with a spirit of innovation, focus, and hard work.

THANK YOU TO ONE AND ALL!

PREFACE

My wife often reminds me, rightfully so, that I spend more hours (waking and sleeping) with electronics than my family. To drive home the point she made a list of all devices. Two (2) television sets, three (3) iPhones, one (1) Kindle, two (2) iPads, two (2) iPods, two (2) laptops, two (2) external hard drives, and one (1) desktop, not counting two (2) work laptops and two (2) Blackberries between the two of us. I have completely discounted the embarrassing number of flash drives, chargers, Bluetooth headsets, and wired and wireless headphones. As funny or strange as it may sound, our household isn't unique in that regard. It is a common situation in most households. We have to admit the fact that we now live in an age where every thought, every conversation, every business idea, every problem and every product has a technology element in it. Whether you realize it or not, you probably cannot go more than five minutes without talking or thinking about a gadget, a system, or a technology in your daily life be it personal or professional. In our personal lives we talk about *technology* and in our professional lives we talk about *IT.* What's the difference? We have come to expect a very high degree of sophistication in products we use in our personal lives. However, except for a very few companies, technology-based products and services built for conducting utilitarian business do not have the same level of sophistication. Examples of Personal Technology product/service providers are Amazon, eBay, Apple, and Expedia. On the other hand, examples of Utilitarian Business Technology product/ service providers are hospitals, banks, insurance companies, and rental car companies.

The difference between personal and utilitarian business technologies is that most business product builders and designers are not day-to-day consumers and users of their own products. This prevents them from understanding the true significance of quality and utility of their products. You cannot be a good cook if you do not eat your own cooking. In our professional lives, we rely on a number of people to help build a product or service for our customers and each person has his or her own perspective, motivation and selfish interest in the degree and type of contribution. This *shared* responsibility often causes production of sub-par products. When individuals are not celebrated, challenged, or fired they have little motivation to strive for perfection. Moreover, most companies tend to focus on Cost, Timing and Quality, in that order, which further leads to Quality getting the short shrift.

This book is called The Golden Age of Drive-Thru IT because we seem to be living in a world where short sighted decisions resulting from quick analysis based on insufficient information, together with quick fire responses, ever decreasing loyalties and lack of relationships seem to be the norm rather than the exception. These habits have now become standard practices and are detrimental to long term financial health of companies just as Drive-Thru restaurants are detrimental to long term personal health and well-being. Even Drive-Thru restaurants are now serving salads and other healthy food choices but IT seems to be going in the opposite direction in its desire to satisfy the insatiable corporate greed of achieving the lowest operational expense structure while charging customers a fee for everything. The term Golden Age sarcastically refers to the widespread acceptance and prevalence of such short sighted behavior.

Current times are close to being the worst, if not the worst, of times for what companies and thus consumers are getting from their utilitarian business technology products and services providers. Companies have resigned to the fact that not much can be expected from IT and that IT needs to be *managed*. IT is treated as a necessary evil that cannot be relied upon to take companies to the next level in their corporate evolution. However, technology departments are not the only ones to blame. Product Managers are their partners in crime. Typical, short sighted behavior of many product managers seems to be achieving parity with what the competition offers. Is it any surprise then that you hear how most *common sense* things you want from your insurance provider, bank, school, telephone service or cable/satellite provider cannot be done because their systems don't allow it?

Technology or IT leaders need to take it upon themselves and prove to the CEO that they can meet high standards for Cost and Quality without sacrificing one for the other. This will happen only if IT takes risks, develops and markets a new image that exhibits professionalism, and believes that its role is significantly more than providing phones and e-mail accounts. IT has to look into its soul, understand dynamics of this digital age and learn to be proactive. All these are possible only by establishing a strategy and, more importantly, executing that strategy day in and day out. Executing the strategy in a cost effective and efficient manner can be accomplished only by establishing minimum thresholds for attitude and aptitude with a singular focus on ensuring everyone understands contributions expected from them. Hiring fewer but smarter people allows companies to strongly row in the same direction and create a cohesive environment that is focused on cost, quality, timing, risks and sustainable profitability.

This book describes various aspects of Technology and how IT can truly rise to every occasion and become a strategic enabler. It describes how IT can become nimble and flexible yet produce robust and graceful solutions that allow companies to drive toward success in an efficient and enriching fashion that CEOs, managers, staff, and customers can be proud of.

PART I

WHAT IS IT?

1 WHAT IS AND IS NOT IN THIS BOOK?

A NUMBER OF BOOKS, articles, web sites and blogs about Information Technology (IT) strategy and management are targeted at C-level managers. However, significantly more non-C-level IT professionals and business professionals are responsible for day-to-day technology and business operations than are Chief Information Officers (CIOs), Chief Technology Officers (CTOs) and other CXOs who, typically, are responsible for higher-level strategies and tasks. This book is targeted at not only those non-C-level professionals but also CXOs and anyone interested in understanding the day-to-day workings of IT and how decisions made daily affect business strategies. This book describes how to connect the proverbial dots in your day to day life, start asking the right questions and start thinking analytically like a true IT professional is expected and required to.

Companies can be divided into three primary categories in the world of technology:

1. Those that produce *ready to ship* technology products that are sold commercially.

2. Those that provide human resources for other companies to build products or produce built to order products.

3. Those that integrate products/services/resources from companies in the first two categories as well as build products for their own use and use by their customers.

This book focuses on IT departments and companies that fall in the third category but all three categories have many elements in common. So, what do I intend to say through this book? More importantly, is there anything left to say that hasn't been said or written before by those much smarter and experienced than me? Most of those smarter people talk about strategies, C-level tactics, grand vision and other "higher level" topics. On the other hand, I am a much simpler person who believes that one has to take baby steps before attempting to undertake grand plans. By no means am I saying there should not be grand plans or grandiose visions. On the contrary, they ARE important. However, focusing only on talking about such plans and visions without understanding a way to execute them is a waste of time. One cannot understand ways to execute such plans without understanding day to day workings of IT organizations. Board rooms and C-level suites cannot expect to solve problems in the trenches or expect grand visions to be carried out with surgical precision without understanding, I repeat, day to day workings of IT. Attempting to apply C-level strategies from the top down has a limited chance of success because making such things work in a sustainable way is like boiling water. How does water boil? Water molecules at the bottom of the container become hot, move to the top by displacing cooler molecules at the top, which move to the bottom, get hot, move to the top and on and on. Similarly, situations in the trenches need to be understood, solved and improved to allow implementation of related Vision and Strategy. This has to be a continual and ongoing process. Vision and Strategy can truly be implemented only at the ground level! Feedback from the ground level needs to bubble up, which might require adjustments to Strategy that would need to be further tested and implemented at the ground level. Vision and Strategy cannot simply be dictated from the top with an expectation that the vision truly trickles down to the bottom-most rung of the corporate ladder.

I set out to write this book with the intention of bringing "trench" topics and issues to the surface. These topics and issues are ones that I have experienced myself, heard about, read about or all of these. I have been employed in or have consulted in the oil and gas industry, a government agency, an educational institution, and financial services industry. I have worked with a wide variety of IT professionals, managers as well as domestic and international clients. These broad and deep experiences as well as my academic background gave me a unique perspective to write this book as I have seen many problems and solutions, failures and accomplishments from close quarters. However, problems and failures seem to repeat themselves many more times than solutions and accomplishments. That trend made me wonder why that was the case. As is human nature, I set out on a quest to find a scapegoat and someone to blame for these shortcomings of IT. As I learned and researched various topics, I realized that I didn't have to look very far to blame someone for these shortcomings. It was IT that created many of the problems for itself. I also realized that it was not as much blame as it was ignorance and lack of self respect and self confidence that allowed challenges to form, fester and repeat. Irrespective of the category of a company, IT departments seem to behave in similar ways and do not look beyond the end of their noses. In other words, IT has not cultivated a habit to think ahead and venture out on its own. IT seems to almost always be content in speaking with its immediate customer. IT is averse to thinking about and interacting with other parties and ultimate customers of their companies. Tell us what to do and how to do it seems to be the prevalent IT mantra and they succumb to pressures of completing projects quickly and cheaply; often for the wrong reasons.

This book is titled The Golden Age of Drive-Thru IT because companies in that (third) category suffer from an insatiable need to deliver results quickly one way or another. Whether the path of delivery is right or wrong, efficient or inefficient seems to matter very little. This attitude seems to be a result of lack of understanding of how IT operates, and the gap between what IT management tells their business customers and the situation in the trenches. As a result, such companies have unrealistic expectations and get addicted to instant gratification of getting an answer on the spot—Yes or No. Either they expect too much or too little and do not realize long term impact of their shortsighted decisions, which are often fueled by insufficient information. This behavior is further complicated by incompetence at various levels of "management" creating a chasm between commitment and delivery.

This book attempts to highlight how such chasms get created, how to prevent them and how IT's true potential can not only solve problems but actually create opportunities and generate revenue. It provides simple yet effective techniques for analysis, communication, risk management and problem solving that help business and IT communicate effectively and efficiently. Clear, concise, timely communication is absolutely critical for success. Communication   cannot be effective if the wrong people (based on talent, attitude, skills and experience) are assigned to projects, tasks, strategies, etc. It is as important to realize how resources are wasted as it is to determine how they should be deployed. Communication failures often result in business and IT blaming each other for the many challenges that companies face. Such factors stare us in the face but we tend to ignore them or think that it is someone else's job to resolve them. More often than not, that someone else does not exist. Technology leaders are often very removed from ground realities and empower their managers to solve such problems and make key decisions. Moreover, *leaders* are typically more worried about protecting their turf and are knee deep in figuring out Strategy that mostly maintains status quo or doing things in ways that transfers risk of action on to someone else. As a result, the heavy lifting of day-to-day decisions falls on line managers or staff. To make matters worse, such personnel are either ill-informed about the impact their decisions are likely to have or they are averse to making big decisions that could burden them with owning up to consequences arising from making decisions. Such behavior is typically at play when you hear someone say, "…that is above my pay grade." That, in turn, results in right decisions not being made and people playing ping pong with critical decisions. This book talks about how to truly empower not only line managers but also all employees so that every decision they make or see being made triggers questions in their minds or explains why certain decisions were made. This collective focus on execution of Vision and Strategy is what will make companies truly successful.

One of the most important aspects of flawless execution is defining "requirements" for various "projects" that are undertaken toward that goal. Requirements and projects have been written in double quotes because they mean different things to different people. Depending on the rung of the corporate ladder on which you are domiciled your understanding and knowledge of Requirements could be vastly different. On one hand, many of us have encountered managers who can talk for hours on end about Requirements but have never written one, never read one nor can they distinguish good Requirements from bad. On the other hand, many of us have also encountered managerial and non-managerial professionals who read Requirements daily, perform their work based on Requirements and are ready to blame Requirements when things go wrong but cannot write Requirements even if their life depended on it. That is why I wanted to discuss Requirements as comprehensively as possible, with due consideration for constraints of space and time. This treatise on Requirements is almost entirely written from the perspective of software development projects. This was so because software-based solutions are such a large part of an organization's IT investment either through proprietary/in-house development, acquisition from third parties or developing them for commercial sale. Moreover, every piece of hardware has software that controls it and commands it. In-house implementations could be custom developed software, implementation of commercially available software, licensing software from application service providers (ASP) or providers in the Cloud. In all cases, determining the true need, the right solution and measurement of success follow a very similar, if not identical, process. Moreover, implementation of non-software based projects can and should mimic process for software implementations and, thus, benefit from a higher project success rate, if done right.

To reiterate, this book is not about any exotic management jargon. This book attempts to describe and solve day-to-day challenges and provides advice on detecting trends to identify opportunities as well as challenges lurking around the corner. It would have served its purpose if almost every, if not every, IT professional can derive benefit from the application of one or more topics discussed in this book. Information contained in this book is not based on management surveys but on discussions with professionals in the trenches whose opinion this author values greatly, together with the author's personal observations and experience. This book attempts to keep presentation of information short and to the point without belaboring a point or trying to meet any size requirements.

Readers should expect to learn how to think of IT as a Revenue or Profit Center with a simple yet powerful integration of Vision, Strategy, and tactical execution. This is a playbook for aspiring IT managers as well as established IT and business executives to encourage new ideas for the new digital world!

2 WHAT IS INFORMATION TECHNOLOGY?

WE TALK ABOUT IT in our daily lives and for most people IT, at work, is personified by the guy who shows up when they need help when their e-mail isn't working or computer is running slowly or there is another such mundane yet mysterious and *life crippling* problem. While we all need that IT guy from time to time, Information Technology is significantly more than that. So, what exactly is IT and what are the building blocks of IT? Is IT a bunch of 0s and 1s, bits and bytes, or a bunch of electrons zooming across the wires commonly referred to as the "network?"

Technology is defined as the branch of knowledge or science that deals with technical aspects in a wide variety of fields ranging from mechanical engineering to making toys to constructing skyscrapers to tailoring. However, technology in today's world immediately makes one think of computers. If only computers and all things digital are considered to be technology then, simply described, the T in IT is different layers of hardware and software stacked appropriately to make information (I) flow from point A to point B whether those points are within the same software system residing on a single physical machine or across the world residing on disparate software and hardware systems. This book also uses this convention of referring to IT as anything that is related to computers, which includes smart phones and wireless devices. The different layers of this stack are listed below. These are broad categories and each category/layer listed below has sub-categories and sub-layers that need to work in harmony and synchronously for a computer to function as intended.

- **Hardware**–a system that is commonly referred to as a server, a machine, a blade or any number of terms but they all refer to the same entity. This hardware system is a vessel that acts as the foundation for all other layers and is primarily comprised of various physical moving parts such as a hard disk, processor chips, fans, etc. that allow electrical signals to be passed back and forth.
- **Operating System**–a base software system (the operating system) that acts as the glue between the hardware system and custom software systems (described below). In most cases, custom built software depends on the operating system on which it is running.
- **Application**–a custom built software system that is commonly referred to as an application (or app) that attempts to solve specific real life challenges by asking questions, crunching numbers and providing responses. However, be forewarned that a base or custom software system is only as smart as the person designing and building it, which is the primary cause of software defects (bugs). More on that topic will be discussed later in the book.
- **Network**–a system that allows *data* to flow from point A to point B (and back, if necessary) whether those points are inside the same physical office or across the world.

A good analogy for this technology stack is the human body. The skeleton, muscles and skin are analogous to *hardware*; brain with all its readily available electrical and chemicals impulses is analogous to the operating system, and individual thoughts, emotions and DNA as the app/custom software system. Nerves, food pipe, wind pipe and blood as the network. Each electronic device has essentially all these layers–be it a $10 TV remote control, a $100 Bluetooth-enabled wireless headset or a $10 million state-of-the-art hardware and software combo. All advancements in the IT world are essentially advancements in one or more of these layers. Someone somewhere is always making an attempt to produce more compact, more independent or more integrated, un-wired devices than their predecessors or competitors did or trying to present a single software suite that can do multiple things simultaneously. For example, wireless or cellular phones used to be just that–phones that allowed mobility–but now phones are mini computers that allow one to navigate the Internet, take pictures and videos, read documents and so on and so forth. This constant quest to produce faster and better products and services (but not always cheaper) have allowed information technology to grow by leaps and bounds in the past 20-25 years.

However, improvements in Application layer are critical in leveraging improvements in Hardware and Operating System layers to create significant, game changing solutions. Figuring out how to leverage technology to create and distribute information, solve real life problems, reasonably predict problems and behaviors, and improve day to day life is what IT is really about. Whether it is developing high yield seeds, developing a 24 megapixel digital camera or enabling communication between earth and the international space station, information technology is what makes it possible. However, these things don't just happen overnight. Usually, a person or a group of individuals envision solutions that solve such complex problems. These individuals are extremely motivated by their goal to achieve and make a contribution than just financial incentives. Similarly, in our daily lives, we come across individuals–in the coffee shop, at the deli, in the adjacent cubicle and even at the DMV–who want to solve problems and help others. They not only think it is their job and that's what they are paid to do but they genuinely like solving problems and helping others. It is an unexplainable quality that people either have or they don't. It is very difficult to acquire or train someone to obtain that quality. Improving professional and personal lives, making this a healthier planet, improving communication, and solving every problem can be a lot of fun when done right. However, just as politicians all over the world are often good at becoming hurdles rather than becoming enablers of solutions, so are many technologists. Either they do not understand technology well enough or are unwilling to take calculated risks for higher rewards. Unless one fails one never learns valuable lessons or appreciates success. The right set of people in conjunction with the right technologies is the only way to attain success that differentiates amateurs from professionals. A willingness to take calculated risks is what allows professionals to aspire for a much higher degree of success than what would be possible with an always-cautious approach and fear of failure. Such attitude starts at the top and IT leadership has to blaze the way. Conventionally, a CIO or a CTO sits atop an IT organization and setting the tone and culture for IT starts with that position. Technology organizations have to be market savvy, technically skilled, organizationally agile, and economically alert to rise above their competition and unlock their true potential. CIOs and CTOs that claim lack of mandate, lack of knowledge, and lack of business sponsorship as reasons for inability to make timely decisions are only taking the easy way out and making excuses. Just as CIOs and CTOs expect their staff to stay on top of advances in the industry, they themselves should get trained, educated, and become aware of business environment so they can be ready to play a leadership role that they are paid to play.

We often hear people say, "I am not good with computers" or something to indicate that they are not technology-savvy. In today's day and age everyone in every company is "technical" whether one likes it or not and whether one realizes it or not. Not having programmed software, built servers, configured network infrastructure or performed any such technical tasks is not an excuse for not enrolling in online classes, reading Wikis, watching videos, or accessing a plethora of resources including community colleges that teach basic technology concepts and their application in various walks of life. It is not being suggested that everyone should learn programming or build servers or write macros. However, everyone should understand how technology works, its role, its typical challenges and strengths, factors to consider in process optimizations, factors to consider in designing solutions, information security considerations, and incorporating strategy into day to day execution. In other words, everyone should be able to relate to almost everyone in the IT organization whether they are database administrators, business analysts, security specialists, project managers, programmers or customer support technicians.

For most CEOs IT is a prime target when it comes to cutting costs. In today's highly electronic world IT is an expensive department to run whether it is due to software/systems licenses or high salaries that knowledge workers demand. However, reducing work force, or as the politically correct phrase goes, right sizing, cannot be the only way to cut costs. If one really thought about IT the right way one can achieve significant and sustainable cost reductions by looking in the right places, reducing costs by increasing automation and not treating IT as strictly a cost center. IT can almost always be a profit center with the right vision, business acumen and true partnerships amongst all business leaders. Leaders have to recognize IT's potential and IT has to market itself in the right way with the ultimate goal of boosting a company's bottom line.

3 PEOPLE, PEOPLE, AND PEOPLE

IT IS HIGHLY LIKELY that you have heard or read that the three pillars of IT are People, Process, and Technology. Some people use another variation—People, Process, Technology, and Infrastructure. No matter which version is used, it implies that if the right people follow the right process and use the right technology then expected results will be delivered with minimal or no unintended consequences. However, companies do not truly take the time to define how those "right people" are identified, which skill sets they should possess, how those right processes get defined and improved, and how the right technology—not necessarily the latest or the most expensive technology—is identified. Who truly has the authority and wherewithal to identify the right mix of all three (or four) components?

When you really think about it "People" is the only component that really matters. You could have a slightly vague or not fully defined process, or a technology that is inefficient (because it is outdated, poorly designed or due to other reasons) but the right people, the smart people can work around those gaps and still deliver results. So, focusing on hiring, and more importantly, retaining the right people should be the most important aspect that IT organizations in conjunction with the Human Resources (HR) department should focus on. HR has to focus on ensuring that every department identifies minimum standards—job skills, attitude and aptitude–to be eligible for hiring. Smart people, who are continually informed about a company's needs and company's strengths, weaknesses, opportunities, and threats (SWOT), can truly be an invaluable asset. Such people have the ability to make course corrections, react on their feet and, in general, do the right things the right way all day, every day. They do not take time away from managers in having to coach them or monitor their tasks and can work well independently or in teams. These seemingly small yet critical decisions pay big dividends that cannot be easily explained or quantified but their effects can be felt and assessed quite easily. When you have the right people the amount of time spent in "managing" them and the amount of time employees spend on "reporting up" can be significantly reduced. Having the right people increases the probability that mistakes don't get made–at least not regularly–and, thus, reducing the time required to coach them, manage them, review them and so on and so forth. People with the right attitude and aptitude do not cause companies to lose time, and, thus, money.

One of the first things that IT organizations have to do is to identify what constitutes "the right team member" and how many team members are needed. A Football team needs Tight Ends, Defensive Linemen, Quarter Back, Receivers, Kicker, Offensive Tackles, and Defensive Tackles and so on and so forth. Football teams define exactly what type of players they need, what is expected from those players and what role they are expected to play. Similarly, IT organizations should state that a certain number of Programmers, Business Analysts, Project Managers, Development Managers, Testers, Network Engineers, Security Specialists, Database Administrators and so on and so forth are needed, at a minimum, to run an effective IT organization. More importantly, IT departments should state that if they get the recommended number of smart team members with the right skills they can execute a certain number of strategies and/or solutions at any given time. Another significant benefit of having smart people is that the overall efficiency of the department becomes very high. In a typical IT department a lot of time is spent on administrative tasks such as time tracking, measurement of (wrong) metrics, status reporting, root cause analysis for every little problem, team meetings to name a few. All these are essentially ways to justify a need for management of all employees due to poor hiring decisions and employees that do not meet "smartness" standards.

Having the right people minimizes overhead–plain and simple. Take into consideration the number of hours we spend in writing reports over and over in multiple formats. Irrespective of the number of hours each of us spends, I am sure most, if not all, of us wish that we could spend fewer hours on such activities and instead spend more time on the real work and letting the work speak for itself. I highly recommend watching a movie called Office Space. If nothing else, it is a highly entertaining fare and makes a few very good points about management in IT organizations. That motion picture is based on a fictional software company called Initech. The Vice President of that company is a character by the name of Bill Lumbergh. Bill's daily routine, in the movie, is to walk around people's cubes to tell them to complete their reports, work weekends or move their cubes. In a nutshell, Bill's character represents a manager who does nothing but tell others what to do and, especially, remind people what they have done wrong. Bill's character also hires two consultants, Bob Porter and Bob Slydell, to help downsize the company and reduce costs. I am quite certain most of us have encountered the character of Bill Lumbergh in our professional careers.

While providing information to one's customers, peers and supervisors is essential for transparency and accountability, doing it for the sake of doing it and doing it the same way it was done for years is neither efficient nor right. There was a standing joke at a company I used to work at. If a person had the title of Manager he/she was automatically assumed to not work or contribute to any solution. Such managers would lovingly call their team members worker bees and never wanted to be bothered by their underlings and managers of such managers never wanted to know what the worker bees were up to or if the worker bees had any problems. Don't get me wrong. Managers play an important role and are essential in organizations but managers have a specific role to play and get paid more to play that role well. They are expected to contribute more in terms of quality and/or type of work. Managers cannot simply be delegators and email forwarders—DEFs, in short. They have to lead from the front and earn their stripes. They have to earn their team members' respect and prove why they were assigned the job to lead the rest of the team and that their job is not to just delegate.

Extrapolating further, retaining the right people results in reduced time doing performance reviews, preparing and monitoring performance improvement plans, hiring and firing, and all such staffing activities. If sufficient time is spent in hiring and retaining the right people then the number of HR personnel required would also reduce just by virtue of reducing the repeated need for talent replacement and replenishment. This leads me to the topic of the (infamous) performance reviews. Most companies say that feedback should be provided throughout the year but actually provide it only annually during the performance review process, if at all. A company my wife used to work for had no performance reviews in her department. Imagine that. Most managers and non-managers heave a sigh of exasperation—usually, for completely opposite reasons—when they get emails from HR on this topic. Most people would agree that annual performance reviews are mostly inefficient, inaccurate, and are quite good at demoralizing people. Such reviews are very subjective, inconsistent and mostly do nothing to encourage or help employees determine how and where to improve/acquire skills. If sufficient time is spent on hiring the right talent then less time will need to be spent on administering personnel. This is another area where HR and IT leaders can draw from the process of software development. The more time you spend up front in defining and identifying the right solution (or person) the less time you will have to spend in maintenance or fixing things on the back end.

If you are an aspiring manager you should realize sooner than later that your primary responsibility is to ensure that you contribute to the team and carry your own weight when you become a manager. Too many team members believe that becoming a manager is their ticket to freedom and that they can just order others around or just boss them around. That is wrong behavior for multiple reasons. For one, becoming a manager is a reward for contributions that are above and beyond as well as different from what the rest of the team brings to the proverbial table. So, upon becoming a manager if the only thought and modus operandi is to find a *resource* to assign a task/project then that is not the definition of a manager.

One of the responsibilities of a manager is to hire and, more importantly, retain talent. Even more importantly, replace personnel when poor hiring decisions are made. Whether you are capable of hiring and retaining talent starts from the job description. Most managers either copy job descriptions that they had previously used or copy one from a job board. While those might serve as base templates, most managers simply tweak a few words here and there and pass them on to their recruiters. This behavior is a classic approach to hiring wherein managers consider hiring decisions to be mundane and boring and unimportant in the grand scheme of things. This could not be furthest from the truth. Just as planning and design are the most important aspects of software development so is the job description in the process of hiring. If a manager does not or cannot take the time to think, plan and write what a team member is expected to do and what he/she is expected to bring to the table how would one communicate that to candidates and evaluate candidates' performance and suitability? Job descriptions should be blue prints for interviews as well as hiring justifications. Simply stating that an extra pair of hands and legs is needed to take on additional workload should almost never be an acceptable justification. Once hired, that job description should serve as the candidate's/ employee's checklist for performance.

Speaking of job descriptions, managers and companies should try to avoid pigeon-holing employees into specific silos. Assigning specific titles of, say, analyst, developer, tester, customer service agent and so on an so forth forces people to be tied to a specific area and breeds territoriality among managers. Moreover, this also prevents interdisciplinary exchange and maturity within the workforce. As employees learn, understand and perform job functions that depend on their core strengths it gives them a better appreciation of their customers—internal and external—and helps them be more productive. Companies can achieve this cross pollination only if they hire smart, logical and analytical people. A smart, logical and analytical person can morph and transform into almost any role, within reason, which provides companies the greatest leverage in adjusting their work force without having to constantly think about hiring in one area and firing in another. In today's age all knowledge workers should be prepared to play at least 2 or 3 different roles and that's how companies can truly create a fluid work force that adjusts to seasonal needs, cyclical needs and, overall, maintain lower cost structures without sacrificing quality. Otherwise, smart people get laid off sooner or later as companies go through the inevitable cycle of alternating growth and downturn. As more and more aspects of the technology world become commoditized and industrialized the only way knowledge workers can survive downturns is by adapting and being prepared with other skills. When faced with unemployment, often, people start seeking out other skills that are *hot* in the market at the time when they are looking for employment and fail to realize that by the time they learn and gain, at least, a working experience that skill might no longer be in demand or that there could be many others with more experience in that skill. Long gone are the days when people could have a core competency in one skill and ride that skill to promotions or into retirement. Knowledge workers have to constantly learn new skills, broaden their horizons and generally have a well rounded view of their companies, industries, and micro- and macro-economic factors.

Just as financial planners recommend diversity in an investment portfolio knowledge workers should diversify their skills portfolio. A programmer can be a Quality Assurance (QA) professional, a QA professional can be a business analyst, a business analyst can be a project manager and so on an so forth. Such acquisitions of secondary skills would also allow knowledge workers to improve their core skills by becoming well-rounded professionals, and learning and improving their ability to evaluate situations from multiple angles.

People are truly the force behind every company's success (and failure). The sooner everyone–leaders, managers, and staff–realizes that, accepts that fact the sooner companies can channel their corporate energy in the right way. What makes people in the company tick, what prevents them from active participation and what would make people get in to the next gear is what leaders should be constantly thinking about and executing on. That can only happen by being closer to them and having a finger on the pulse of their teams and departments. If you are a manager you should ask yourself do you know every member of your team and does every member of your team know who you are? If you are not a manager what have you done to reach out to your manager and your manager's manager so they can hear and appreciate your ideas, challenges, and concerns?

4 HOW TO THINK ABOUT IT

WHEN YOU LOOK AT a house what do you most admire or get turned off by? Features and characteristics that you interact with daily and have day to day utilitarian as well as aesthetic value will affect and attract you one way or another. It is highly likely that furniture, wall colors, lighting, floor plan, kitchen, and bathrooms and such features that interact with your senses greatly affect you. It is less likely that you get attracted to or turned off by the garage door, lock on the front door, light switches, and number of doors. Various aspects of IT can be viewed the same way.

Information Security is like locks and security alarm systems. Every house needs them and is expected to have them but one rarely pays attention to them. IT infrastructure is like walls, windows, doors, door frames, etc. Every house needs them and is expected to have them but one rarely pays a lot of attention to them as long as they are in the places where they should be and do the basic things they are expected to do. However, business software is like wall colors, type and placement of furniture, amenities in the kitchen and bathroom to name a few. It is highly unlikely that you think about the type of door when locking it on your way out but the right amenities in the bathroom and kitchen are important and provide the little joys in life and make your home more enjoyable. We adjust bath water temperature depending upon season and weather. We adjust lighting and position of furniture depending on what we are watching and who we are watching TV with. Similarly, business software/software used by employees to conduct business should have significant focus from technology and business leaders. That is not to say that Information Security and Technology Infrastructure should not receive due focus but business software is always front and center for customers and users. As a result, there should be significant focus on design, aesthetics, and other visual elements as well as the number crunching aspects of the software system. Those should not be an afterthought and technology leaders have to lead the charge on that front. If the software is poorly developed or implemented then it results in frustration and employees will use it only because they have to and, more importantly, just enough to stay out of trouble. Employees should *want* to use software that the organization is building or buying for them to make money for the company. Business executives have to have their finger on the pulse of their employees' thinking and thoughts about technology they are expected to use in order to be efficient and productive. In this day and age companies cannot afford to have business executives who do not have a good handle on technology and its capabilities or at least have the ability to challenge their technology leaders. Companies without such savvy executives cannot expect to be top tier companies of the future. When executive managers and heads of various departments do not challenge or demand bigger and better things from their technology product and service providers a chasm starts forming between what companies can achieve and what they actually achieve. Of course, cost is a significant factor in all decisions but there usually is a cheaper way to do things and an expensive way to achieve those same goals. The key is in finding the right solution for the right situation at the right time for the right price.

Given time and money technology can do almost anything. When people tell me that something cannot be done it only means, in most cases, that they have not exhausted all options of solving that problem or they have no clue how to solve it. There seldom are situations wherein technology cannot solve a problem. When such situations do occur there usually are alternate ways of solving those problems. Solving problems the right way always requires problems to be defined in a clear and concise way. Co-mingling problem definition and proposed solution(s) at the outset makes it difficult to define, explain and solve problems as parameters of the problem definition change over time. IT leaders and managers should challenge themselves every time someone says, "this cannot be done." IT and business leaders have to move away from asking "can this be done" and get in the habit of saying "let's do this." Every conversation about business opportunities and challenges should consider not only cost, timing and risk but also creativity and innovation that would have a positive impact on customers–internal and external. After all, perception is reality! And the Wow factor is highly underrated.

Most companies still think of technology systems as a cut-and-dry math problem. They focus on system design from the perspective of moving data from one point to another or displaying data on a screen or a report. While data is critical, a customer's interaction with the system is equally critical. A system could be exceptionally built in terms of data processing, error handling, scalability, etc. but if the customer interaction (a.k.a. user experience or UX) is poor then no one will use the system or people will use it because they have no choice and then curse IT under their breath. The system will be dead-on-arrival. Would the iPhone have been as popular and user friendly if it looked like a pager or if a feature phone looked like a brick? Your guess is as good as mine.

So, what should business leaders demand and expect from IT? More importantly, how should IT raise the bar for itself in terms of performance and perception? Business leaders have to demand creative, innovative solutions and have IT provide facts and opinions on costs, risks, time lines, and various options to achieve the end result. There are almost always multiple options and IT should make it a standard practice to present all options. Thereafter, it should be a collaborative decision to determine returns on investment. However, if IT expects business managers to define exactly what they want and how they want it i.e., detailed Business Requirements then it is a big problem. Both IT and business managers have to think about tools and systems and "processes" they use in their personal lives and the real life business problems they are trying to solve or market opportunities they are trying to capture. Problem definition in its most raw and nascent form is the real key to achieving success and stellar growth. Often, IT and business managers tend to focus too much and too soon on the *solution* instead of the problem. Let me elaborate. One business manager I used to work with never used to define the problem correctly and would jump headlong into defining the solution precisely how he wanted it–flush with database design. Not necessarily the right design but a design, nevertheless. He would get mad if anyone asked him for a problem statement. He would say that he knew what he wanted and it was IT's job to implement his ideas–some were quite good, others very questionable. His philosophy was–don't ask questions. Just do it. He personified the new business manager–one who knows not only his business but also enough of technology. In some cases such managers know just enough technology to be over-confident in what they think they know, which can be very dangerous.

A couple of factors have changed over the years in terms of technology awareness amongst business personnel and technology product providers. (1) IT awareness among mid-level business managers within the conventional profit centers has increased considerably over the years. That is a very good thing. (2) With increasing commoditization of technology products and increasing availability of solutions in the Cloud, it is becoming easier for sales people from such technology providers to throw around buzz words with such enthusiasm that you'd think they were promising to solve world hunger. A combination of both these factors results in many solutions being purchased and adopted without due consideration for obvious and potential pitfalls as well as long term costs. The philosophy seems to have become very short sighted with little or no consideration for long term impact on cost and health of the company. A few years ago a very large company wanted to sell a product that provided ability to *write* "IF/THEN/ELSE conditions" in their most basic form. Nothing more, nothing less. It was not even branded a rules engine. Moreover, those IF/THEN/ELSE conditions were not integrated with any other solutions. All those *sophisticated* features for a measly $500,000.00 toward setup/implementation plus $25,000.00 monthly toward system maintenance. The sales person, rather account executive, and the product manager reached out to one of the business managers with this proposal. The business manager, in a somewhat typical manner, started socializing it within the organization and was absolutely convinced that this product would solve many, if not all, of his operational problems. Long story short, that product was not purchased for obvious reasons but this is a classic example of how easily certain personnel can get hooked on to the next shiny thing peddled by small and big providers alike. Why did this happen? Primary reasons were that IT was not aware of that manager's needs, that manager did not trust IT, and that manager did not believe that IT would provide a solution.

What IT can and should deliver and what IT actually delivers could be almost identical if managers in charge of IT and managers in charge of business both leveraged each others' strengths and trusted one another. They should also constructively challenge each other by putting themselves in each others' shoes as well as their ultimate customers' shoes. Once the problem/opportunity is defined the right way and at least some of the parameters such as time to market, cost appetite and risk tolerance are well defined it becomes easier to determine the right solution. May be it starts out as a simple, small, inexpensive solution with the flexibility to grow into a more robust solution that is easy yet not too expensive to maintain. More often than not problems are not well defined and parameters are not well set. As a result, solutions seem over engineered, too expensive or simply miss the mark and then complaints abound. Thereafter, the system and process becomes an anchor that drags a company down. The knee-jerk reaction then, in most cases, is to throw good money after bad to "fix" the bugs. Experience and research have shown that ongoing/annual cost to maintain an IT system is, typically, 50%-80% of cost of developing the system. The exact percentage depends on how well the system was originally designed and developed and how maintenance costs are computed but, needless to say, maintenance costs are a big component of total cost. However, most IT professionals do not consider this during the design and analysis phases. They are too eager to design and develop based on requirements provided to them without thinking long and hard and asking difficult questions. They think they can always come back and fix things, not if but, when they break.

Total cost of ownership (TCO) has to be a factor in decision-making in everything related to technology. License costs, ability to rebuild parts of a system, potential down time–scheduled and unscheduled, number of people needed to oversee a system, configurability, and flexibility have to be considered in the TCO. TCO is an extremely good yardstick to measure a smart and visionary solution provider. A visionary solution provider will not only provide a solution that solves for short term parameters of need, cost and risk but also builds the necessary building blocks to be able to extend the solution in anticipation of growing or changing needs. One doesn't need to be a business expert or have a crystal ball to know what people might want and expect from software as users of that system grow more comfortable and more sophisticated in their daily interactions with the system. One doesn't need to be a *rocket scientist* to figure that out. One just needs to understand what consumerization means and how that plays out in day to day life. Consumerization refers to the trend that new products and services first appear in consumer markets and, consequently, businesses are forced to adopt them due to customer as well as employee demands.

Consumerization of IT has become a very powerful concept and IT has to be very aware of its impact. As employees and customers are exposed to a wide variety of systems, interfaces, UXs they take the best parts of each such system and expect those to be made available in all systems. When they don't, guess what happens? IT's image takes a hit, trust takes a beating and everyone loses. Leaders, managers, strategists should be very aware of consumer trends and efficiency tools available in the marketplace. Consumer (personal) technologies are often introduced by new entrants and entrepreneurs who are hungry and eager to introduce *disruptive* products and technologies whereas business IT is often flat-footed and very risk-averse. Start-ups and young companies have little to lose in terms of reputation and market share and larger corporations run the real risk of reputation loss, regulatory scrutiny, public opinion backlash, etc. but that is no excuse to sit back and say No to everything. Companies have to strike a balance between risk mitigation (or avoidance) and being on the bleeding edge.

Part II

NUTS AND BOLTS OF IT

5 REQUIREMENTS AND IMPLEMENTATION

IT IS VERY EASY to say that companies should strike a balance between risk and innovation but why do so many companies continue to do more of the same? Why do so many companies get stuck in a technology rut or go out of business (after they achieve a certain scale)? A big part of this problem lies in the approach to execution and assignment of responsibilities. Let's review how most companies approach the process of execution a.k.a. Requirements and Software Development Life Cycle.

Whenever my wife and I decide to go out for dinner we are seldom certain of our destination. We always end up with an exchange similar to the one below.

"Where do you want to go today?"

"I don't know."

"What kind of cuisine are you in the mood for?"

"Thai or may be Italian but definitely not Indian or Mexican."

"Okay. Thai or Italian it is."

"Let's pick a place that won't have a long wait and the food is good."

"And, by the way, I don't want to drive for more than 15-20 minutes. I had a long day at work and I am very tired."

This is a quintessential requirements gathering conversation. In this specific case my wife is the *business owner* and I am the *analyst* trying to determine where I need to drive her and achieve the *project goal*. Every project starts with requirements but how many IT managers really understand what requirements should contain and what is expected from requirements and how much time should be spent on requirements? On the other hand, most IT managers expect their business unit counterparts to provide requirements in excruciating detail and are very quick to point fingers at their business counterparts when projects get delayed or scope changes creep in. Very few line managers understand what requirements should contain and even fewer higher level managers understand that. One has to understand the sorry state of requirements and the process guiding it to understand why such a large number of projects fail or do not provide the intended benefit. Not acceptable but understandable.

Every IT project follows one version or another of a structured implementation ladder also known as software methodology. These implementation ladders or methodologies have the following steps— vision and scope, requirements, design, development/configuration, documentation, testing, training, and deployment. Certain methodologies go through all steps multiple times while others go through each step only once. Moreover, depending on the methodology more or less time is spent on each step. However, irrespective of the methodology requirements are the backbone of every project/implementation since all other steps are hinged on the robustness of requirements, rather problem definition.

The process of completing requirements timely and, more importantly, accurately is what determines the probability of a project's success.

Information in this chapter is relevant to companies that build software for their internal use or as a product to be sold to other companies. In other words, it applies to all companies and departments whose existence is based on building software or ordering customized software from third parties. Moreover, one project's requirements are usually not leveraged for other related projects. It is no wonder then that mistakes get repeated, which leads to a standard response from management that they will ensure mistakes are not repeated and that long term fixes are implemented.

Let's say you are asked to provide requirements for a toy for children 12 months and younger. Would you just state that the toy make a certain sound and, say, flip over every time a button is pushed or would you also provide guidance on weight, height/length, color, type of material, etc.? This is where software requirements fall short. They almost never focus on all aspects of what it takes to create a professional quality product. For instance, requirements might focus on user experience but not on avoiding "bulkiness" or "unwieldiness" of the final product. As a result, when flaws due to poor design surface in the form of high maintenance costs, performance bottlenecks, lack of flexibility, etc. guess what happens? A new project is created with requirements to be written by the same people who wrote the original requirements and programmed by the same people who created the original problem. And the cycle continues.

Requirements can never be completed as one phase as prescribed in the Waterfall methodology neither should they be completed piece-meal as prescribed in the Agile or Scrum methodology. Establishing Requirements is really a discipline and not just a task in a project plan. Requirements should be developed by considering current business needs, lessons learned from previous implementations, and potential future needs. A Requirements document is the result of analyzing the problem from multiple angles. Depending upon the type of organization and project/solution there could a few more or less factors but most projects would, typically, need to consider the following factors.

- Data Quality and Quantity–existing data, if any, as well as new data created by the software
- Customer experience
- Functional business goals
- Alignment with business and corporate Strategy
- Technical implementation
- Information security
- Initial as well as ongoing maintenance costs a.k.a. total cost of ownership/operation (TCO)
- Training needs–to ensure that the solution is intuitive, easy to use and intensive/ongoing training is not required
- Barriers to user adoption
- Time sensitivity of the required solution and
- Constraints and Limitations

Without considering all these factors each project could take an organization further and further in a direction diametrically opposite from its strategic direction. The worst part is that this would happen right under the noses of executive management with management being none the wiser. Ensuring that the process and discipline of creating Requirements for IT is repeatable and reproducible is paramount for the entire company. Upfront investment of time and diligence in creating Requirements automatically translates to lower risk and cost in downstream steps of technical analysis, programming, testing, deployment and rollout, user adoption, and ongoing maintenance. Otherwise, companies focus on saving a dollar today only to spend ten times as much in ongoing care and feeding of the system. Creating a repeatable, reproducible and robust process also prevents dependence on individuals' interpretations and differences in opinion, perception, and other emotional and professional inconsistencies that exist amongst various team members. The importance of preventing such dependence dictates that Requirements should be created using templates and checklists that are customized to serve an organization's specific needs. Additionally, these templates and checklists should be supplemented by documents/plans that describe business and corporate Strategy. These

Strategy documents should be available at all times as references to all stakeholders and actors involved in execution of these strategies. This approach ensures repeatable, consistent execution of Strategy to maximize efficiency and probability of achieving business goals.

The TeChSt Method

Every organization will need many templates. These templates can only be established by investing time and diligence from the right team members but once established they can pay huge dividends. This method of building/writing/creating requirements, being introduced in this book, is called TeChSt (pronounced, Text), which is an acronym for **Te**mplates, **Ch**ecklists and **St**rategy.

Templates allow organizations and IT departments to consistently implement features and functions using standardized criteria across the enterprise rather than leaving them to the discretion of individual project managers and project teams. Even if IT managers are disciplined enough, management changes, outsourcing and delegation of implementation to inexperienced team members can result in lack of consistency. Even if only one project perpetrates a different methodology that one project is enough to create the cracks for every other project and an organization's Strategy to falter. Along with these templates (explained in detail later in this chapter), each set of requirements should also draw from checklists that ensure that all tasks needing completion are indeed completed and, with time, allow perfection through repeatability and reproducibility. With *doing more with less* becoming every organization's mantra and cost, competitive, and information security pressures at the forefront the constraints and challenges facing technology teams are changing by the day. Not having to rely on individual memories has never been more important. Moreover, we no longer live in the age where people would join a company and stayed loyal till their retirement. We are in the golden age of at-will employment. No single person's continuance at a company can be taken for granted. A person working at a company one morning could be gone by the end of the day or the end of the notice period. Additionally, this is further complicated by the fact that when technology leaders have rightsizing or downsizing as the only lever for reducing costs they invariably lose institutional knowledge and intellectual property through firings or voluntary resignations. Such personnel losses, irrespective of the reason, further lodge an organization in that ether of mediocrity, if they are not prepared for such eventualities. Checklists are a simple and easy way to achieve Six Sigma- or Total Quality Management-like (TQM) efficiencies without having to spend hundreds of thousands of dollars in training, certification and other avoidable spending. Such simple implementations also reduce cost of training and ongoing maintenance by not having to relearn lessons learned in the past and avoiding mistakes typical in an inexperienced or undisciplined organization.

Strategy documents are those elusive documents in an organization that probably very few people get to see. More importantly, people that are responsible for ensuring strategies are implemented via ongoing projects would probably have never seen such documents, assuming they exist. However, strategy is always on the agenda at various management and committee meetings but attempts to connect the dots with day-to-day operations are rare. Questions about execution of such strategies are also equally rare. This is not to say that none of the companies do this but I am willing to bet that the number of companies that do not far outweighs the ones that do.

This TeChSt method of implementing projects has another significant benefit that most regulated companies can relate to—audits. Whether internal audits, external audits or examiners from regulatory agencies are considered, having a consistent, easily explainable and repeatable process can pay huge dividends during audit exams. It is surprising that despite knowing what auditors examine and like to see IT managers do not pay attention to addressing those needs at the outset. Conventional, albeit incorrect, thinking seems to be doing whatever is necessary to reach the next milestone as soon as possible without considering the long term costs—quantitative and qualitative—of such short sightedness. Consider the classic example of building custom software for which timelines are laid out even before the scope of the project is defined. That is just human nature. Irrespective of software, hardware, home plumbing or an automobile project the first two questions we ask are—how much will it cost and when will it be done? We often fail to provide all the necessary details so we can get an accurate estimate but when the actual time and dollar costs are more than the estimate we all ask why the estimates were incorrect? Well, the simple answer is that the estimates were, in all likelihood, correct but *scope description* and *scope containment* were poor. When IT gets burnt by such situations the knee-jerk reaction is to become very conservative in estimating projects, which leads to accusation of "sand bagging." IT professionals need to better understand human psychology and human instincts. This will help them predict questions and their reactions. That, in turn, leads to a better understanding of your customers, a higher level of mutual trust, and a higher quality of truly valuable deliverables.

Template Types

Let's further discuss templates in the TeChSt methodology. Each template focuses on a specific aspect of Requirements and provides guidance for completing routine tasks, which would allow more time to be spent on high value tasks that would differentiate the product and/ or service from its competitors. Each organization should have at least the following templates:

Communication Standards Template (**CST**; for both Internal Customers and External Customers)

> This template should provide guidance on pre-implementation as well as post-implementation communication. For example, how internal and external customers should be informed about upcoming changes, how soon should they be informed, how

detailed the communication should be, what kind of examples should be provided, how will the system provide automated status updates and/or reports, if applicable, who should be contacted in case of problems and so on and so forth. Such templates should also ensure that relevant communication includes factors considered before implementing the solution and why certain factors were considered and others, if any, were ignored. When people get to know what is being done and why you are likely to get a much higher buy-in than if you just forced it down someone's throat and said, "This is what you now have to do. Trust me this is for the company's good and we don't care what you think or whether you like it."

User Interface and Interaction Standards Template (UIST)

This template should define:

- How web sites and graphical user interfaces (a.k.a. GUIs, UIs or front ends) should interact with users of a system?
- What type of data should be collected?
- What level of data granularity should be collected?

Examples of type of standardization that can be achieved are manual entry of dates or a visual calendar, display error and validation messages per data field or as one comprehensive list, placement of those messages, font to be used for ease of readability and multi-browser, multi-device consistency, accessibility of help and frequently asked questions (FAQs), navigation style—roll over, fixed, or slide down, and so on and so forth.

Another important aspect of UI standards is to know cardinal rules of building a UI. Many professionals, especially those who build the guts of a system, do not pay attention to building a UI because they think UI building is not very sexy or intellectual work and then there are those that do nothing but build UIs. If you ask me, building UIs is important. I would draw an analogy with food and cocktails. When we go to a nice dine-in restaurant and order, say, lobster do we expect it to be just placed on a plastic plate and some garnish thrown around it or do we expect nice crystal ware with garnish tastefully decorated around it in a visually pleasing presentation? Or when you order a Mojito do you expect mint leaves, sugar, rum and ice mixed together and thrown into any old tumbler or do you expect it in a highball glass with mint leaves garnish? The answer is obvious and the reason being that visual presentation is important. It is said that we consume as much with our eyes as with our mouth. Similarly, building a simple yet clean and functionally rich UI is critical in customer attraction and retention. If users don't find the UI rich, appealing and inviting they are unlikely to use it at all. Or if they use it once they are unlikely to be repeat customers/users.

Technology Interfaces Standards Template (TIST)

This template should guide which programming languages, infrastructure products, security protocols, authentication and authorization mechanisms, and encryption algorithms will be used. This template should also define whether software systems should be built using two-tier or n-tier architecture, what kind of business logic should be in each layer and whether databases should contain business logic or if they should serve strictly as data stores. Type of data to be collected from customers can be either in this template or the UIST. More importantly, though, personnel responsible for data mining, analytics, determining trends and benchmarking should have a say in the type of data collected during each interaction with the customer. Otherwise, the downstream post sale analysis will be ineffective and yield inadequate/undesirable results, at best.

Data Interfaces Standards Template (DIST)

This template should guide how data models should be constructed and how data should be exchanged with business partners and customers, where appropriate. It should also guide whether one or more formats could be accepted and/or distributed or whether a single standard should be adhered to. Depending on a company's role in the partnership a company might be able to enforce a data interface standard upon other parties or a company might have to adopt an industry standard or another company's standard but this problem can be approached in several ways. One way could be to maintain a standard data model and interface protocol internally and then convert it to a partner's format just before transmission. This allows the data model and information architecture to remain pristine within the company and helps maintain a single standard across the enterprise. At the same time, it also helps conduct business with partners in a timely manner without losing business for the sake of sticking to principles. After all, the primary goal of IT is to help conduct business in an efficient and effective manner.

Maintenance Standards Template (MST)

This template should define the following:

- How ongoing maintenance should be conducted?
- Who should review the type of maintenance being performed?
- How maintenance items should be prioritized?
- How customers and end users reporting these maintenance problems should be notified?
- How customers and business owners should be notified of proactive maintenance tasks?
- Previous maintenance items completed, lessons learned, preventive steps implemented, etc.

The last point is very critical. One of the goals of an effective and efficient IT department should be to proactively monitor its hardware, software, routers, switches, circuits and all such infrastructure and operational components. Without a standard, well-defined protocol for performing routine maintenance problems will always be solved reactively and almost never prevented. Moreover, one of the worst reflections on an IT organization's ill-preparedness is when customers report problems and IT is blissfully unaware. When problems are reported IT promptly gets to work by opening a ticket, getting people on a conference call, followed by multiple questions in attempts to find out who to blame or in attempts to prove it is not their fault. After a long, painful process IT fixes the problem but what follows next is really *interesting*, to say the least. Typical responses are:

- A previous software/security patch or firmware update caused the problem due to insufficient testing (implies it was the testers' fault).
- A file did not transmit correctly (implies it was the business partner's fault).
- We changed a network setting (implies not sure whose fault it is)
- We updated a security policy (implies someone higher up or a business partner told us to do this).

The classic conclusion to all such problems is…. We will ensure it never happens again! While not all problems can be foreseen, it is a *standard protocol* for IT managers to prepare a resolution summary stating what the problem was, what the resolution was and why it wouldn't happen again. It is imperative that IT provide transparency into problems and solutions but IT should state and business customers should demand to know if those problems could have prevented. If so, why weren't they prevented? Managers and leaders are paid to prevent problems and not just react after the fact.

Many of the problems are predictable and numerous tools are available that help prevent many of the problems but without a plan or awareness of business operations IT organizations can only react and will seldom be prepared to prevent problems from occurring. Without realizing what exactly was previously done and why it would be very difficult to train new team members and ensure that veterans do not get careless and repeat mistakes.

Process and Technology Backup Standards Template (BST)

Process and technology backup and preparedness primarily apply to all companies that are directly or indirectly involved in financial transactions, health and human services or other such industries where data loss is not an option. Some amount of unavailability or loss of data within a small window of time is acceptable depending upon the industry but not having a plan is not an option.

As a matter of best practice many companies have a plan but not much else. These plans are, usually, called Business Recovery or Business Continuity Plans. Companies, typically, have plans if they are required by their regulating agencies, Board of Directors, business clients or if it makes for good public relations. However, very few companies are really prepared to handle emergencies. If a real emergency were to strike a vast majority of these companies would not be able to start conducting business as fast as their plans state they would. If companies with C-level risk management officers and government committees and sub-committees were not able to predict, foresee and/or act upon the mortgage-related financial crisis of 2007, how can we expect them to handle a real, physical emergency that could come with unpredictable forces and full fury of nature? Performing a thorough test and simulating an emergency is critical to a successful verification of an emergency plan. Without such simulations it would be impossible to get all pieces working together for the first time when half the people are likely to be unreachable and the other half won't be thinking straight. More importantly, such tests will cause implicit assumptions to bubble up to the surface. For example, availability of electricity, cell phone service is an implied assumption but how many plans explicitly state such assumptions? After super storm Sandy some neighborhoods in New York and New Jersey had no power for weeks whereas most personal emergency preparedness plans call for storage of food and water supplies to last 3 days. Learning from Sandy and what a massive earthquake could do personal plans need to be revised and tested to ensure you can actually get to your emergency kit, and know how to use it. Similarly, corporate emergency plans need to be constantly reviewed, revised and tested not once but multiple times each year.

Solutions Management Template (SMT)

A Solutions Management Template should be the mother of all templates and encompass inclusion of all other templates. A SMT should be the first template to be pulled out of the quiver to determine which other templates apply. SMTs can apply not only to technology solutions but also business analysis, market research, financial analysis and almost every type of solution or research. Most research, analysis and technology initiatives do not leverage previously completed work. This is due in part that knowledge bases do not exist, partly because people do not realize the importance, partly because institutional knowledge is lost (over time), and partly because people are undisciplined. Irrespective of the reasons, not spending time today in preparing for future needs implies that every day companies and individuals fall further and further behind and inertial forces get stronger and stronger. It is similar to following a diet and exercise regimen. With every passing week the task of eating healthily and exercising regularly seems more and more difficult and people eventually give up or it becomes a case of too little too late. A Solutions Management Template should, in conventional terms, be thought of as a Project Plan. It can provide guidance on types of teams to engage, types of factors to consider, strategies to consider, risk tolerance, legal issues, regulatory considerations, budgets, and timelines. Without a master aggregation template every individual team and template could only do so much and the key to ultimate success lies in seamless and flawless integration of individual plans.

All templates have to work in conjunction with checklists for there always are curve balls in the business world. If companies spend most of their time in mundane, routine tasks they will never be prepared to handle those curve balls and hit them out of the park. The importance of templates and checklists becomes evident as they start yielding incremental benefit with every (successful and unsuccessful) "project" completion. It is a self-fueling machine. In other words, the more it is used and applied the more benefits companies derive, which prompts them to use it further.

Importance of Checklists

Checklists provide a simple yet extremely effective way to ensure that everything that needed to be done was indeed done. Checklists are used by airline pilots, car mechanics, manufacturing plant managers and a number of other professionals in a wide variety of industries. So, why don't we—knowledge workers—use them? One reason could be that the IT industry is not yet mature and disciplined enough in its processes. IT industry is a baby compared to airlines, manufacturing, and automobile industries. IT professionals are cowboys, in some ways, and unless these cowboys are guided by an IT industry standard the power of knowledge and discipline cannot be fully tapped and unleashed on the business world. As an example, everyone has come to realize, even though no one says it out loud, that defects/bugs/issues are just a fact of everyday life in IT. What works one day is not guaranteed to work the next day. Granted that a lot of factors and variables affect the functioning and performance of software but each such factor and variable can be understood, documented, controlled, planned for back-up and so on and so forth.

Consider a typical data center. A data center consists of routers, switches, servers, cables, software applications, databases, etc. How many organizations would know the type of impact and where it would be felt if a switch or a router stops functioning or needs to be upgraded or *patched*? Your guess is as good as mine but it is safe to assume that that number is likely to be small, very small. Of course, most data centers have redundancy to avoid single points of failure but the point I am trying to make is that in some ways IT organizations fly blind and have multiple blind spots. If an organization were to maintain a checklist of all the components connected to each component and have an inventory of all such daisy chains it would dramatically reduce the number of outages and disruptions and those that cannot be prevented can be closely monitored. That way, such disruptions can be fixed before end customers are unpleasantly greeted by "Error. Your request cannot be processed." IT staff can be coached to leverage such checklists for every change. These checklists can be in the form of printed pieces of paper, spreadsheets or databases and can be formed by leveraging knowledge bases or repositories of collective experience and lessons learned. IT staff can be fully brought on-board and the resistance to changing their standard operating procedures can be eliminated by telling them one simple fact. The more they leverage such knowledge bases and checklists the fewer times will they need to be "on call," disturb their personal/family time and the fewer times will they be needed to work on unplanned schedules. Following a checklist is in their best interest - plain and simple. However, before everyone starts building checklists and appointing Chief Checklist Officers there needs to be careful planning and determination of types of checklists required in an organization, how they will be maintained and, most importantly, an organizational commitment to diligently build them, use them and keep them current. Otherwise, it would become a case of garbage in, garbage out and more garbage in. I cannot stress enough the importance of keeping information current without which they would become outdated in the blink of an eye. I have seen far too many projects, systems, tools and checklists rendered useless because organizations and individuals were not committed to keeping them updated.

An easy way of keeping such checklists updated and current is to make the task of updating them contextual. In other words, integrate such tasks as steps in a person's daily job duties and existing processes. If such tasks are kept as independent steps, not integrated with daily processes then it begins to feel like a chore to complete them and we all know how diligent most of us are about doing our chores. As stated earlier, these templates will help IT departments and, in turn, organizations run business operations like a streamlined, well-oiled machine. Templates in conjunction with checklists will help IT deliver incredible value to its internal customers and, directly or indirectly, to its end customers. Without such simple yet essential tools the cost of building, buying or licensing software will keep growing exponentially and stakeholders should point as many fingers at themselves as they would like to point at others. There comes a point in every company's growth where the cost of maintenance is as high as, if not higher than, investments needed to create value-added products and services. At least a basic understanding of how software is built, how it is maintained and costs involved will help every stakeholder gain a better appreciation for IT and learning how to work with IT in ways that help the entire company. Without this understanding business managers will continue to take digs at IT and IT will continue to find itself in a corner trying to defend its turf sometimes transparently and sometimes through mumbo jumbo.

IT has to aggressively work on minimizing the amount of time it spends on maintaining status quo or baby-sitting technology components. IT has to get busy figuring out its contribution to the bottom line by active involvement in business decisions and strategy or forever be relegated to being a cost center. As they say, charity begins at home. If IT doesn't believe it can contribute to the company's profitability then no one else will.

6 BUILDING SOFTWARE

So, NOW YOU HAVE wonderful Templates, Checklists and other processes and tools to help significantly minimize the amount of time spent on maintenance and "bug fixes." Now what? Now, it's time to build value and spend as much time as possible in furthering strategic goals.

Let's say you have an ideal set of requirements that all stakeholders, including IT, worked together to craft. Now, what? Well, it must be time to throw it over the wall for the programmers next door or half way across the globe to start coding, correct? Well, not so fast. Let's think about it a little more. What are the expectations from any software? The answer really depends on who is answering that question, who's building the software, how it is planned to be used now and in the foreseeable future, and how it is being paid for.

Many aspects are common to the business of software development no matter who is developing it or how it is being developed. Building software takes time and money—sometimes, a lot of time, and, thus, a lot of money. Upon completion, typically referred to as deployment or releasing to production that software needs to be maintained even if it is not being enhanced. This is because, as described earlier in this book, business software is a layer sitting on top of other layers and changes to one or more of the variables that make up those layers could adversely affect that software. Moreover, software, no matter how flexible and configurable, is built on certain assumptions. A single violation of an assumption results in unintended consequences and undesirable results. As a result, software needs to be monitored and maintained. Moreover, most software systems have data being fed to them from other systems or through human data entry. This could also affect the functioning of the system. Let's face it, a software system is only as smart as the person who designed it and will work flawlessly as long as none of the assumptions are violated and none of the links in the chain are weakened or broken.

Software could also break every now and then due to one or more virtual or physical moving parts in its functional supply chain. Business owners/Customers will require software to be enhanced from time to time and sometimes too often. On a lighter note, if they did not do that half of the IT professionals would be out of work–job security, indeed! Under certain circumstances software will seemingly behave erratically. Business owners will call it *broken* and IT will call it *working as designed*. Often business stakeholders seem exasperated with responses from IT and IT jargon. Avoiding this communication gap between IT and its customers requires IT to improve its communication and explain steps involved in software development, steps involved in isolation and identification of potential problems, and steps taken to resolve problems.

Business owners can provide requirements and mandate that software be built with a great degree of precision. As an example, they can specify UI colors and fonts, whether an event should be triggered with a single click or double click of a mouse, what the screen navigation should be and so on and so forth. Such precise definitions/requirements are wonderful but results in a colossal waste of time when time is spent in defining the same aspects for each project/solution. Imagine the amount of time a company spends when multiple business unit managers/staff spend time specifying the same aspects over and over. Chances are that either the company, as a whole, is performing the same tasks over and over or different approaches are taken for each project/solution resulting in a disparate and disconnected customer experience. Investing in building software with such precision is very expensive in terms of time and money spent in programming and testing. Such common aspects should be extricated as templates and checklists that could be reused with a high degree of certainty. Smart companies can go even a step further and build components that act as software templates. In other words, companies can build plug-and-play components or libraries that can be dragged and dropped into other solutions. This approach is not unique or ground-breaking. Companies that develop operating systems and programming languages already provide such libraries as part of their Software Development Kit (SDK). Business technologists have to leverage these kinds of tactics in their execution and save their companies valuable time and improve operational efficiencies.

As a result of *templatization* of such routine tasks IT architects, technology managers and other personnel can focus on the true challenges that various real life situations present. Every project–large or small–has its challenges and it is up to those leading the projects to recognize them, disclose them, address them, and prevent them from becoming permanent and chronic problems. Challenges in building large scale software systems increase exponentially due to a significantly higher number of moving parts and larger number of people involved. However, such challenges and risks can be mitigated by understanding typical problems encountered at each step and proactively avoiding them rather than reacting to them after they occur. Treating a large project as a number of smaller projects can further help you be proactive than reactive.

Challenges in implementing large scale software systems depend upon the type of project–new development, migration from a legacy to a new system, migration from a proprietary system to a third party/vendor system or first time implementation as a hosted third party system. Challenges vary from one implementation phase to another and are unique to each phase. Some of the most critical and recurring challenges that are typically encountered have been discussed here.

Design Phase

One of the primary challenges of designing software is defining Design. How do you define Design? Ask yourself, ask your manager, ask your peers and ask your customers what software design is. You will be surprised to hear the sheer variety of answers you will get. Why is that? Design is different things to different people and applies differently to different solutions/problems. At a company where I used to work, the head of software development had mandated that every change (large scale new development, minor defect fixes, and everything in between) had to go through a design phase before development could be initiated. However, there was no definition of design or guidance on what should be considered in design or what the outcome of a design phase should be. Design is nothing but a blue print that provides clear guidance on what's acceptable, what's definitely not acceptable and what can be left to the discretion of the implementation team. Think of it as the blue print of a home that a city's planning department reviews and approves but doesn't really care if the inside walls are painted, white, blood red, yellow or black. Design documents/guidelines, very much like corporate and business strategy documents should be always available to the implementation teams and be used as a reference at all times. It, however, does not mean that the design documents/guidelines get cast in stone. On the contrary, design guidelines should be reviewed periodically and any updates should be applied uniformly to all applicable systems.

Four (4) types of Design components need to be considered—Software and Security Design, Hardware and Infrastructure Design, Validation Design, and Data Design. Each type of design is, typically, completed—explicitly or implicitly—for every solution implementation. Each type of design is critical in its own right. Without careful consideration of each type of design the entire design, and thus the system, could be flawed since even a single weak link could bring a system to its knees at the wrong time.

Software Design should consider building blocks of business logic and security governing the software. Building blocks of business logic should determine flexibility and configurability of the software with an eye toward predictable needs and long term maintenance. Architects and designers should think of software as a virtual home—a home where every occupant's activities are continuously tracked and monitored. Tracked not to snoop or watch over their shoulders but to ensure that they do not trip and fall. However, if they do fall that fall is promptly recorded and notified to the right personnel. You should need and want to know if someone trips and falls in your home. You should not have to wait to hear from the person who fell down. For example, software systems should be designed to track unanticipated and anticipated errors so they can be promptly monitored to determine causes and solutions, if necessary. Recreating problems is an expensive proposition and the best time to capture a problem is precisely when it occurs. This would allow subsequent notification to users as to why they encountered an error and why the system misbehaved or which assumption was violated. The worst statement a software designer and developer can make is, "I do not know what happened. I cannot recreate the problem." Software systems are very predictable and they will behave the same way each and every time as long as the same parameters and variables are presented to the system.

Security Design should consider if and which parts of the system should be built as separate modules in order to be administered discretely. Depending on the type of business purpose there are likely to be authorization models and limits that would need to be considered for strong process and security controls to prevent abuse and fraud. Such design should also consider whether security controls should be applied at the software layer, hardware layer or both.

Hardware and Infrastructure Design should consider whether software should be built to handle increasing and decreasing capacity using distributed processing or compounded processing. This is also referred to as horizontal scaling and vertical scaling wherein processing capacity is increased by adjusting number of machines in the processing pool or by adjusting capacity of fewer machines that are capable of significant processing power.

Within Data Design establishment of a data model is a step that's often ignored early only to be regretted later. Data models are critical in serving immediate business needs yet allowing sufficient extensibility to address future needs. However, there is a fine line between over-engineering and providing flexibility. Given a choice every business manager would state that everything should be configurable. Just in case. However, flexibility and configurability have an associated cost. Flexibility is inversely proportional to time, cost and maintenance overhead. Only true business technologists can help determine the right balance of quality, amount and retention of data. Data is truly the life blood of every organization and it has to be afforded its due importance in development of solutions. Otherwise, companies end up accumulating an inordinate amount of data without knowing what to do with it or even knowing what kind of data they have and how much data they have. In other words, companies accumulate huge virtual landfills of data that consume significant system and human resource costs by way of data storage and backups.

Like requirements, design can never be completed as a single phase. A framework of guiding principles, DOs and DON'Ts, key decisions can be established before development begins but software designers/architects cannot stay away from development. Too often those that have the responsibility to design either have no experience in development of software or consider their job completed by the time development starts. Design decisions will almost always run in to roadblocks and challenges during the development phase. At that time, if designers/architects are not actively engaged in understanding and solving problems then the purity of design gets compromised even before the system is fully developed. Such roadblocks to design will also be encountered during testing and subsequent enhancements after the initial deployment. This happens because during testing and real life usage a lot of unanticipated scenarios crop up due to lack of sufficient planning or changes in market conditions and assumptions. In such situations there is no choice but to solve for those roadblocks while ensuring design remains robust and can withstand the test of time.

Such unanticipated situations that arise, due to insufficient planning or changes in variables, necessitate architects to be thoroughly engaged throughout the development lifecycle. Another suggestion for architects is to be hands on by building prototypes of systems, which can be used to prove a theory or viability of a system. It keeps architects on their toes in terms of staying current with technology and potential problems that might occur during development. It also allows IT to visually showcase a concept to their customers and business counterparts without having to over-verbalize rather poorly verbalize the solution because we all know how good IT is at communicating new concepts and ideas to the layman. Prototyping and software modeling prevents designers and architects from sitting on a high perch and dictating design without having any skin in the game. They cannot treat design as a mini-project or a task in a project. Architects have to be held responsible for ensuring their design is technically robust and secure, economically sound and flexible enough to withstand the test of time.

The importance of various design criteria depends upon the end use of the software. For example, software that customers use to purchase products or perform financial transactions has to be highly scalable, available and extremely high performing. On the other hand, consider software that is used to process a batch of transactions or exchange data with business partners and, say, becomes "active" only once at the end of each business day. Such software has to be highly available and scalable but performance is not critical as long as it finishes its processing within a reasonable amount of time. Software that interacts with end customers has to be dynamically scalable using the concept of virtualization and continuous monitoring in the right way to assign or de-assign capacity depending upon usage and volume. Availability and performance are a function of sound architecture, building reusable components that have specialized functions and components that can be easily unplugged and replaced with better ones.

Development Phase

As discussed in the previous section, during the development phase key architects tend to take a passive and/or reactive role. As a result, it becomes a challenge to ensure that software being developed adheres to all design principles and that band-aids are not being applied to reach the milestone of completing development on time. The other challenge during development is interpretation of requirements. This challenge is especially true in a Waterfall methodology wherein developers and business analysts or business operations managers are not in constant communication.

Developers are notorious for making assumptions and not being cautious. They tend to assume that the requirements and design provided to them have been well thought out and that their job is to simply follow the orders. As an example, when developers are provided requirements in which certain words are accidentally misspelled they will copy the text verbatim without reading the text, informing someone or correcting the text. Copy and paste is the mantra most developers follow, which, sometimes, leads to disastrous consequences. An executive, at one of my former employers, once very casually made a statement in an all-staff meeting that was the result of this quintessential problem. He said, "… our software has ticking time bombs." To provide you with a little more background, one of the programmers had incorrectly programmed a calculation and that calculation had been baked into the system for over 5 years. However, that calculation never got evaluated since it came into play only under very specific circumstances. Such a circumstance either did not occur or went undetected for almost 5 years. When that circumstance did occur it caused a mini disaster that caused the IT department to scramble for damage control and resolution came almost 3 days later. However, that one instance was enough for the IT department to be seen in bad light for a few months. No jobs were lost over it but it caused enough disruption that some of us still remember it as egg on IT's face.

When that disaster occurred what was the IT team's response? We implemented per requirements and testing did not catch the problem. Those may be technically accurate reactions but one look at the requirements should have told the programmer, and anyone else who read the requirements that it was a problem waiting to happen. In other words, a ticking time bomb, as the business executive so eloquently put it. Developers have to take responsibility for their programming actions and perform their due diligence before handing off their code to testers. Developers should certify their code and not play the game of whack-a-mole wherein they fix defects if and when they are identified. Developers and IT organizations have to make zero-defects a goal for their development teams. While it is difficult, it is not impossible. Changes will always come up during the testing phase but there is a huge difference between defective code and updates due to changes in assumptions or requirements. IT should consider creating incentive programs for meeting goals such as "zero defects." After all, what encourages a sales organization can also encourage a technology organization. Such cultural changes are essential to change the mindset leaders and *followers* alike, which will result in the necessary transition to a Profit Center.

Testing Phase

Execution of testing should be based on the outcome of Validation Design. Two broad types of test categories should be considered– Functional Tests and Regression Tests. Both types of tests should be based on Software and Security Design, Hardware and Infrastructure Design, Data Design, and Validation Design. Functional Tests are those that validate new changes/features. Each new change/feature should be thoroughly validated. Testing should focus not only on first time usage of a system but also on seasoned data. In other words, testing should focus not only on information imported or entered for the first time but also information that sits and seasons or marinates in the system. As time passes, many systems have pre-built logic that results in, say, interest accrual on a Savings Account, late payment on a loan, expiration of a marketing offer, maturity of a contractual offer and so on and so forth. All such situations should be considered during validation design. Without such detailed considerations systems are likely to have ticking time bombs that seemingly blow up when least expected.

Functional Testing should follow a four step test cycle (think of it as four rounds of tests).

1. Key test cases are provided to the development team to validate the programming. Passage of these tests should be part of the certification proposed in the previous section.

2. All new features are tested by the Software Quality Assurance team. At the end of this Step, all tests should be assigned a Successful or Unsuccessful status.

3. If some tests are found to be unsuccessful then the broken features should be fixed and only those features should be retested in Step 3.

4. If Step 3 is successful then all tests performed in Step 2 should be repeated. This ensures that none of the features that were previously working (during Step 1 and Step 2) were broken as part of Step 3 test fixes.

You could utilize a 2-step or 3-step approach if the complexity of the system/change is relatively low but, ideally, a four (4) step approach should be utilized.

Once functional testing is completed regression testing should be undertaken. Depending upon timelines, system complexities and other factors execution of regression testing could overlap functional testing.

Regression tests are those that verify all features of a system that existed even prior to introduction of new features. Typically, two sets of Regression Tests should be executed–Full and Limited. Full Regression Test Set is one where every feature–simple, complex, critical, non-critical, etc.–is tested. A Limited Regression Test Set is one where only features that are critical to business processes are tested.

Every system upgrade should follow Functional Testing and, at least, Limited Regression Testing. Whether Limited or Full Regression Testing is undertaken depends on the nature of features being introduced and whether they affect existing features. Typically, IT and business managers should collectively determine which type of regression testing should be executed. Complexity of programming, business process impact, financial impact, regulatory risk, number of users, and types of users impacted are some of the factors that should be considered in determining the right type of regression testing to be conducted. Great, you successfully completed functional and full regression testing for one project. Now, here comes the second project. Can you reuse the same regression test cases? Not really. Each project's functional test cases have to feed the regression test cases to keep the regression test cases updated and current. Otherwise, testing becomes an unplanned, undisciplined process causing numerous problems that result in finger pointing and time lost in fixing issues instead of working on another productive project. It is best to understand and avoid problems that typically arise during the functional testing phase.

The biggest challenge during the functional testing phase is to ensure that correcting one defect does not introduce other defects or break previously corrected defects. This is a common occurrence. Sufficient comments within the code and relevant documentation help minimize, if not eliminate, such occurrences. Developers, typically, have a very narrow and localized focus when fixing defects. This is partly because they may or may not be aware of the big picture impact and software might have, originally, been developed by another programmer/company. The other reason is that testing is often rushed and performed with a compressed timeline. As a result, the quality of fixes takes a backseat to the time taken for fixing them. People tend to forget that it is better to take a little extra time and fix defects right the first time. Instead, managers focus on "defect seasoning" by counting the number of days (or sometimes, even hours) that a defect has remained open. When you put an incorrect amount and incorrect type of pressure you are likely to get undesirable results. This leads to waxing and waning of defect counts. I have seen people play games with status reports wherein they mark defects as Fixed just before a status report is generated. This makes them "look good" by increasing the number of defects in their bucket lower than those in the testing team's bucket. This is the result of IT managers', intentionally or unintentional, pitting of the testing team against the development team. Implementing a zero-defect policy helps mitigate such gaming of the process. Just the fact that a defect is found—as long as it is a true defect and not a change in requirements—should be counted as a shortcoming of the   development team. On the other hand, if no defects are found then the development should deserve a reward/commission similar to a sales or customer service team achieving a predetermined goal.

The other testing challenge that companies face is due to incomplete or complete lack of performance and stress testing. Performance and Stress Testing allow a system to be tested under excessive levels of stress and strain in attempts to determine breaking points. Such stresses and strains could result from a large number of users concurrently using a system, large number of transactions flooding a system, or other infrastructure bottlenecks. Remember, business software is comprised of a daisy chain of multiple components and the overall customer experience is only as good as that of the slowest component. Failure to perform such testing or performing it inadequately results in unforeseen problems resulting in loss of revenue and reputational risk. Examples of incorrect or incomplete testing tactics are:

- Incorrect transaction volume assumptions.
- Incorrect peak volume or concurrent volume assumptions.

- Testing on better or worse hardware platform than that used in the live environment.
- Assumption that the real world has remained static since the "project" began.

So, how do you avoid these pitfalls without spending an inordinate amount of time on repeatable testing tasks? Many commercially and freely available tools allow testing to be automated so development teams can make necessary adjustments for optimal performance and features. Testing for performance is one of the most important steps especially when the software is used by a large number of customers and processing speed is critical. Moreover, hiring the right, smart programmers ensures that they program in the most efficient way. All programming styles and strategies are not equal. Efficient ways of writing code are as prevalent as poor ways of writing code.

Without a disciplined approach and rigorous process performance issues tend to creep in slowly and eventually they manifest themselves through a seemingly innocuous change. That innocuous change becomes the last straw that breaks the camel's back. Then the finger pointing begins. IT says the change was so simple that it could not have caused a performance impact. Customers say the system is running slower than it used to and IT should just fix it. Well, guess what. Nothing happens. Everyone goes back to their day jobs till someone remembers or decides to report the problem again and the "conversation" continues…

We need to figure out root cause.

We need to fix it now.

May be the problem is in the network.

May be security policies changed.

May be a patch caused it.

May be we should revert the change and so on and so forth.

Deployment and Rollout Phase

This phase is by far the most challenging. While one would like to breathe a sigh of relief after months, if not years, of hard work this phase is the litmus test of all the work and investment up to that point. While ensuring that software is properly packaged and deployed to production remains a challenge in itself, the real challenge during large scale software implementations is server to server communication. Production and non-production servers, usually, are "located" on separate network segments to create separation between real/live data and test data. This separation results in differences in security configuration, server configuration, and the level of access developers/testers and other systems have. To complicate the situation even further each application tier (web tier, application tier and database tier) is often developed by a different team and teams, sometimes, make assumptions that lead to situations where one tier does not communicate effectively with another tier even though the same code seemed to have worked in a pre-production/test environment.

Although their intentions are good, developers have a bad habit of granting themselves (or having someone else grant them) the highest level of system access possible in a non-production environment. On the other hand, developers and systems they develop have level of access that is absolutely necessary in production environments. Thanks in large part to justifiable paranoia of security and network gate keepers. What happens then? Developers say it worked in test but I don't understand why it doesn't work in production. Duh!

The second type of challenge, and a more serious one, is user adoption and measuring success of the implementation. Software development often focuses on delivering functionality requested and there is limited focus on ensuring that the investment is utilized as expected. People are inherently resistant to change irrespective of how good that change might be. As a result, it is extremely critical to determine usage and adoption and have a plan to eliminate processes as well as systems, if any, that the new system is replacing.

This should be done by developing a "marketing plan" for systems rollout. You can use traditional marketing and viral marketing techniques but take your systems on a road show. Talk them up. Don't be afraid to set high expectations amongst users of the systems. In other words, try not to be overly cautious and risk averse to the point where IT and your systems are perceived as just a big BLAH! You have to generate a certain level of hype and then deliver to match that hype. You have to pique interest among prospective users of the system and leave them wanting to learn more. IT can learn lessons from the movie industry by showing users previews and trailers of the soon-to-come system rather than send them emails and training material just before it is launched. Conventional ways of providing information to users about new systems is not only a bad idea because user input is critical but it also makes people feel that they never had a say in the development and configuration of the system. People are much more persuadable and acceptable of change when they know they had a hand in its creation and configuration.

Post Rollout Follow-up

Every large scale implementation has subsequent phases that are a result of scope creep and/or deferred defects and enhancements. Implementation of such changes, invariably, results in alteration of the original design and/or layering additional logic in the system. Lack of documentation and insufficient code comments make it extremely difficult to remember and determine why certain things were implemented a certain way and which options were considered before various decisions were made. Additionally, after rollout of large systems development fatigue sets in and IT personnel become flat footed, take vacations or move on to other projects. In other words, the A team is not around or the team doesn't bring it's A game to the drawing board. As a result, mistakes get made if those bug fixes or enhancements invalidate previous decisions/logical design. Such errors can be prevented by a laser-like focus on adding copious code comments and preparing documentation. Documentation does not automatically imply reams and reams of paper that no one ever reads. Instead it should be a cheat sheet of sorts or a check list that helps ensure the original design is maintained and if business logic is implemented in a unique way then the next programmer that tries to modify it or debug it should not have a hard time figuring it out. Think of a good business presentation. It contains talking points that a speaker elaborates on but the talking points clearly indicate the highlights of what the speaker plans to convey.

Moreover, documentation should be contextual in nature. It should be accessible when and where needed. That is to say it should be accessible from within the code. IT has to get smarter and current with times by recording videos and adding copious comments within code or linking to documents that might be stored elsewhere. After all, we live in a hyperlinked world. Get used to it and make use of it. The days of traditional documentation in the form of documents and spreadsheets are long gone even though most organizations continue to dwell in those practices. They are not the only ways to document decisions, policies or procedures. Usage of video, audio and traditional documents as a complement will create a richer and, more importantly, more efficient experience for everyone. This is another example of how IT should shed its conventional thinking to learn newer tricks to improve itself and, ultimately, lives of its users and, thus, the company's bottom line.

So, what are the three most important architectural principles in building software?

a) Reusability of components to provide agility to the business model.

b) Robust data model/library that includes an authentication and authorization mode.

c) Loosely coupled components to allow a plug-and-play model with easy upgrade or redesign of individual components, as necessary.

7 POWERING UP

You will find as many variations of IT organization structures as flavors of ice cream. However, just as all flavors of ice cream have sugar, milk and eggs as common ingredients, most companies have one manager at the top with more managers below that one manager and the pyramid extends further down. Furthermore, most companies have many common elements such as C-level titles of CEO, COO, CFO, CRO, CIO, CTO, CMO, and CAO. Roles of CIO and/or CTO typically report to the Chief Operations Officer, Chief Financial Officer or the Chief Administrative Officer. This is primarily due to the fact that IT is seen as a department that provides tactical support, a necessary evil and expense, and a department that needs to be managed and administered as an overhead. None of these are right ways to look at an IT organization. If one does not realize the potential that their IT organization can deliver then they either have the wrong skill sets in the organization (both inside and outside IT) or they do not need a formal IT organization. Most mid-sized and large companies, however, do need fully functional and well staffed IT departments because IT can be a strategic enabler. However, what that really means and how it can be an integral part of a company's strategy is difficult for most people to comprehend let alone plan, implement and execute. As discussed in previous chapters, conventional thinking should not be applied to organizing IT departments. Individuals that make up an IT department should be carefully stitched together to allow maximization of every individual's strengths and minimization of weaknesses and inexperience. It is simple math with a twist of gut instincts and character judgments. The power of IT is the ratio of combined strengths in the numerator and combined weaknesses in the denominator. The only way to maximize this ratio, and thus the power of IT, is to maximize the numerator and minimize the denominator. Both have to be simultaneously achieved to truly power up IT and allow it to fire on all cylinders.

Why, in so many organizations, do CTOs and CIOs report up to CFOs or COOs? The reason is simple, very simple. People with the power to structure organizations do not understand the true value of IT. When IT *leaders* themselves do not understand the value that IT can provide how can they explain it to others? The President of a company where I used to work once told me that when times get tough most companies shut down their marketing departments (that company did not have a very big IT department) to save money but he believed that during such times companies should market their products and services even more aggressively. Companies with sizable IT departments should think in a similar manner. While most companies try to reduce costs by reducing people, smart companies can be opportunistic and invest further in automation of the right processes and systems to reduce long term costs to overcome tough business climates. When times are tough most companies tend to reduce spending and invest less in technology-based solutions. Most companies think that the only way to manage IT costs is to cut costs by lowering headcount in the IT department. Reducing staff is an effective way to reduce costs in the short term but without planning and executing plans to sustain efficiency companies will be stuck in the rut of hire-fire-hire-fire. When times are tough companies should think creatively about the tasks and processes that are most labor intensive and automate them to reduce dependence on a traditional operational workforce.

All companies have to manage costs and managing IT costs is usually a challenge since costs can add up very quickly and technology evolves constantly. However, IT cost management does not mean not leveraging technology to improve efficiency and productivity. Technology and customer service are two distinguishing factors that help differentiate a company and truly become its customers' partner. Technology can help achieve the following:

- Create innovative products and solutions.
- Lower cost of doing business.
- Increase productivity.
- Reduce risk.
- Improve flexibility and consistency.
- Allow creation of proactive as well as detective controls.
- Enable timely communication.
- Provide real time information among other benefits.

On the other hand, IT in the wrong hands can create bottlenecks, excuses, unnecessary expenses, frustrations, lost opportunities and, above all, increased costs. This may be a reason why IT is typically labeled a cost center when it can actually be a significant profit center. It can generate profits and improve the bottom line if only technology leaders can identify its true potential, assign the custodians and know how to measure its successes. Otherwise, IT becomes just an order taker and implements projects from incomplete and poor problem definitions—and an expense black hole that keeps sucking in dollars year after year with little to show in terms of true return on investment (ROI).

What can be done to ensure that a company—its management and employees—do not have an incorrect perception of IT? I recommend two primary actions—appoint business technologists as heads of the department and have them mentor and guide all other key managers and employees. A business technologist is a person who understands technology, its benefits, pitfalls, costs and returns while also understanding a company's core business and has the vision to appropriately and creatively apply technology to improve the lives of staff and customers. Of course, a single business technologist or even a group of business technologists cannot offer a silver bullet and need to work hand in glove with smart business professionals from other departments such as marketing, sales, operations, and finance. However, a business technologist has to either envision and create solutions or convert someone else's vision into reality in the most efficient and effective manner. A business technologist should be expected and trained to know when to build a solution, when to buy one and when to say that a manual process might be best employed. That business technologist's job should be on the line if right decisions are not made. He/she should also be willing to stand the ground and be passionate about IT and not resort to provide-me-requirements-and-my-team-will-execute mode.

Establishing the right organization structure is critical to ensure that IT gets the right representation, is leveraged effectively and measured accurately. Otherwise, technology-based business solutions just become half-baked, under-utilized and an excuse for finger pointing. That, then, leads to the perception that IT does not deliver value and that it is just a necessary evil. Companies need to do two things to solve such *artificial* barriers and challenges. They are artificial because they are a result of lack of planning, analysis and knowledge. The two things that a company has to do are to establish the right organization structure, and understand human behavioral trends/patterns as they apply to using various technologies. While an IT organization has a high impact on the effectiveness of projects the beneficiary business units also have a corresponding impact on a project's success. Typically, a corporate COO and/or individual business unit COOs have overall responsibility for running a company's business operations but these traditional business roles do not do justice to companies operating in today's high tech world. This role should be split into a Business Assembly Officer (BAO) and a Technology Adoption Officer (TAO). Each business unit should have both these roles and these two positions together should be running the day-to-day operations of a business. Technology solutions are such an integral part of almost every company's operations that is makes no sense that a manager sitting in a *cost center* is expected to provide help to the business unit whenever necessary. It is like asking a person living in Washington D.C. to create laws to solve every need of people, say, living in Biloxi, Mississippi by understanding their lives, habits and needs. That kind of strategy and thinking will not yield the right results. Organizations have to realize that a massive amount of money is spent on IT with little to no return on that investment and definitely not as fast as it could or should be.

So, what are a BAO and a TAO and how are they different from a COO? A BAO should be responsible for managing business operations and assembly of various business components and functions such as offering the right products to the right customers, providing excellent customer service, ensuring compliance with all company standards/ policies and laws of the land. On the other hand, a TAO should be responsible for ensuring the right toolbox is created, delivered and kept working at all times. The toolbox I am referring to is nothing but the right set of software and hardware systems to help:

- Improve productivity
- Lower task completion times
- Increase consistency
- Ensure 100% adoption of the right tools in the toolbox
- Identify and eliminate inefficiencies
- Envision solutions that can be communicated easily and effectively

Often times, either a traditional COO does not have the necessary technology background, skill set, experience or he/she is pulled into so many different directions that focusing on technology based solutions and strategy gets de-prioritized and delegated to a line manager who neither has the vision nor the motivation. In other words, the job gets delegated without sufficient guidance, direction or marching orders and these managers keep doing *more of the same*. This is why a TAO with the right caliber, seniority, focus and skills is absolutely essential to running a successful and efficient business operation.

Let's discuss the overall structure of an IT organization and how a TAO and a BAO fit in. Organizations typically have a CTO or CIO sitting at the top of the pyramid and managers reporting to the CTO/CIO are assigned carve-outs of the IT pie. This is in many ways similar to how politicians carve out electoral districts for political gains. These managers then have other mid-level managers and these mid-managers have analysts, developers, and testers in their teams. Such structures results in at least five layers between the head of IT and the person putting the widgets together. With each layer there is more and more loss of information due to translation or *summarization* and in many cases information simply does not travel from top to bottom or bottom to top making the two ends of the organizational spectrum completely disconnected.

Although organizational structures will vary depending upon type of industry, type of company, type of executive leadership and types of systems supported and developed, many common elements should be considered and factored into every organizational structure. An IT organization at the very top should be led by a Chief Technology Strategy Officer (CTSO), and a Chief Technology Adoption Officer (CTAO). From a corporate hierarchy perspective, a CTSO should have two direct reports—a Chief Data Officer (CDO), and a Chief Information Security Officer (CISO) while a CTAO should have two direct reports—a Chief Software Officer (CSO) and a Chief Technology Operations Officer (CTOO). These 6 key executive managers should be responsible for collectively running the IT organization. While having six executive managers, instead of just one, might seem like additional bureaucracy and layers, the point I am trying to make is that there are many areas that today's technology teams need to focus on. Depending upon the size of a company more than one responsibility could be assigned to an individual but that individual should be aware of specific areas he/she needs to focus on and not make it one big blob of responsibilities called *IT Management.*

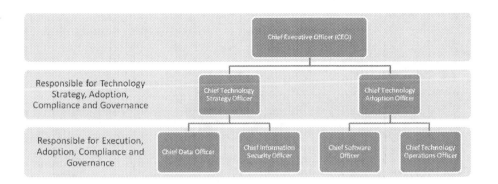

Technology Leadership Organization

Chief Technology Strategy Officer (CTSO)

A CTSO should be entrenched in the business Strategy to understand the vision and roadmap of business operations laid out by the CEO and the business unit leadership. A CTSO should be in a position to provide strategic recommendations and course corrections depending upon availability of technology solutions, cost of technology implementations, partnerships with other companies, and market trends. A CTSO should have the necessary foresight, organizational authority and professional courage to take calculated risks to help a company get ahead of its competition and achieving strategic goals. A CTSO's focus should be as much outward looking as it should be inward looking. Without being in touch with goings-on in the technology space as well as industry of that company's core business, continual education, and customer trends and behaviors IT departments and their leadership end up becoming the frogs in a well. Reliance on consultants with "industry expertise and experience" becomes order of the day. A CTSO should aspire to become an indispensable, trustworthy asset to the CEO and business unit leaders by virtue of his or her business and technology acumen.

Chief Technology Adoption Officer (CTAO)

A CTAO should be responsible for ensuring that right technologies are made available and adoption remains high. A CTAO should be the execution arm of IT's strategic initiatives. A CTAO should make the right technologies available by way of creating an effective, efficient combination of software and hardware configurations. Specifically, in a software products company, a CTAO should have a very good handle on how its products affect and improve their clients' business processes and affect personnel using them. It is, relatively, easy to develop a product and many companies can do so. What is really more valuable is to show a BEFORE and AFTER comparison of conducting business with and without that product in the software landscape. CTAOs should be able to explain the How and prove the Why without limiting themselves to presentations and technical mumbo jumbo. CTSOs and CTAOs should be in constant communication with BAOs and TAOs to ensure that situations on the ground are addressed timely and effectively.

Chief Data Officer (CDO)

A CDO's role is extremely critical and becoming very challenging yet interesting in this day and age. Chief Data Officers should be to a company's data what Chief Accounting Officers and Chief Financial Officers are to a company's finances. This role is that critical. Companies generate enormous amounts of data but seldom do they have a strategy to handle it. It is analogous to growing landfills due to discarded consumer products. Companies are very focused in increasing production, acquiring new customers, generating reports, selling services or sharing data with partners and so on and so forth without realizing or defining a life cycle for data. Although most companies recognize importance of data and information, very few companies invest the time, resources and money that data management deserves. One key difference between physical landfills and data is that physical landfills become another country's problem or government's problem but data remains a problem of each and every company doing business. These virtual landfills continue to grow within an organization and there is no single person responsible for *waste management.*

A CDO's responsibility should be creation and maintenance of a data balance sheet. It should provide insights into:

- Sources of New Data (e.g., Customers, Business Partners, Internal Processes)
- Channels of New Data (e.g., E-mail, Web Services, File Transfers)
- Data Consumers (e.g., Executive Management, Business Analysts, Customers)
- Data Destinations (e.g., Business Partners, Regulatory Agencies, Customers)
- Data Manipulators (e.g., Data Analysts, Report Writers, Data Profilers)

Companies primarily generate data from selling their products and services and logging detailed transaction-level data for each such sale. What happens to that data from the time it enters a company's systems and how it traverses from one point to the next or who uses it remains a mystery to a great extent. Solving this mystery typically takes a large number of IT and business unit personnel. Moreover, such exercises are conducted on an as-needed basis and results of any on-demand exercises become unreliable soon after the completion of such exercises. More often than not, even after spending a considerable amount of time people will be left scratching their heads because the mystery would not be completely solved. Moreover, the speed with which businesses like to move and the pace with which information evolves creates a mind boggling situation that will make anyone's head spin. Add that to the fact that most businesses have departments that act as disparate, disconnected mini companies and the causes leading to data hell become crystal clear. Note that the problem itself or possible solutions do not necessarily become clear but the reasons for the mess do. The road to cleaning up such a mess is long and hard but it does lead to a solution. A team led by a knowledgeable CDO has to take on this responsibility to reach data salvation.

CDOs should hire and train smart analysts that truly know how to read into data sets and look at them the right way. When you look at it in the right way data talks, sings, and dances in an attempt to tell a story. You just need to keep your mind, eyes and ears open to see and hear that story. The role of such data analysts is going to grow in importance as companies try to get their arms around the massive amounts of data they have been and will continue to collect. It would be in companies' best interests to start hiring and cultivating that skill now than waiting till it becomes the in thing to do. CDOs should not give too much weight to prospective employees' previous titles. Titles such as Data Analyst, Data Modeler, and Data Architect do not automatically imply expertise in data research, analysis, and design.

A pictorial view of the data problem has been shown below and will be used to discuss various short and long term solutions.

As shown in the picture labeled Data Sources and Dissemination, Customers result in creation of new data, Business Partners feed new data and Employees create new data, typically, via e-mails, spreadsheets, databases and documents. New data is consumed by multiple departments such as Finance, Accounting, Sales, Operations and Marketing. Invariably, each department creates a copy of this data because each department looks at a different *cut* to arrive at different conclusions. What happens to each copy of this data? It gets dumped in the landfill! A CDO has to track and organize every single piece of data that enters a company's systems and how it traverses from one system to another. Once sources of data entry are tracked each piece of data has to be tightly controlled to prevent it from mutating and multiplying. The biggest culprits of these data mutations and multiplications are employees in various departments that do not like to share and work off a common data set. They are like spoiled brats who do not like to play in a common playground. They like to carve out little territories to create fiefdoms and managers either do not realize or do not care that their employees are being inefficient and creating long term problems. It is considered someone else's job and that someone else, typically, does not exist or is completely ineffective due to lack of authority, staff and budget, or blissful ignorance.

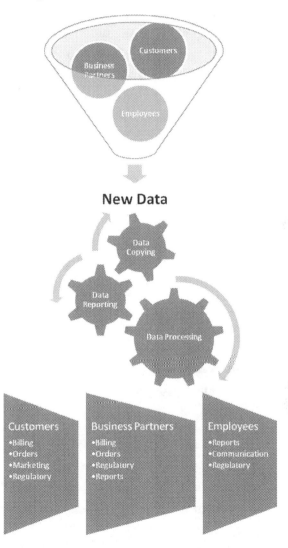

Data Sources and Dissemination

A CDO's number one job, with full authority, support and backing of the CEO, CTSO and CTAO, has to be tracking, controlling, disseminating and dismantling of data at the right points in time. In other words, a CDO has to create a life cycle for each type of data collected and created. Certain pieces of data could have a thirty-day life cycle while others might last for twelve years but each piece of data should have well defined start and end dates. Data cannot forever remain in a state of suspended animation. The essence of that data should be retained, if required, in the form of a decision, a summary or an aggregation. It is said that the devil is in the details. A different twist on that phrase is that the devil in these details will kill your data strategy. Get rid of unnecessary details at the appropriate time. This data can be transactional data about a sale, an e-mail, a spreadsheet, a memo or a digitized version of an interesting magazine article. Along with clearly defining start and end states, a CDO should also define containers such as a database or another storage mechanism where each type of data should reside. For instance, data related to a sale would be copied to a reporting database 2 hours after the sale and reside in the Sales system for 90 days. After those 90 days it would reside only in a reporting database for 5 years. After 5 years that data would be summarized and <u>moved</u> to another reporting database and then deleted after another 5 years. In this example, data would reside in only one database for almost all its life except for the first 90 days. The assumption is that during the first 90 days there could be customer service needs such as returns and exchanges that would necessitate availability of data in more than one system.

There should be systematic restrictions and controls on how and where data gets stored, copied and moved. Without intricate knowledge of data and specific controls Social Security Number, ATM PIN, Credit Card Number, Phone Number, Account Number, Marital Status and Amount of Last Payment could all receive the same treatment. Without a robust storage strategy data management becomes extremely difficult and chances of data loss and data theft increase exponentially. This is one of the primary reasons why we hear of cases about customer information leaked or stolen from all types of industries. This problem   is   not limited to any one specific industry. However, the bigger a company the higher the chances that it will be a target of outside hackers or foul play resulting from social engineering. Chances increase exponentially if a company operates in financial services or healthcare industries due to significant amounts of personal and detailed customer information collected.

One of the reasons why I propose this group of 6 leaders be responsible for IT becomes very clear when we discuss the role of a Chief Information Security Officer (CISO) and how a CDO's responsibility partially overlaps with that of a CISO. A committee typically has a bad connotation with an undertone of rampant bureaucracy but these 6 leaders have to manage and operate IT by fashioning a nimble, forward thinking, security-oriented, business friendly committee. One might think terms such as security-oriented and business friendly are contradictory to one another but that would be an incorrect assumption.

Chief Information Security Officer (CISO)

A CISO's job description is very simple but the actual job is very difficult. Very simply stated, a CISO's job is to Prevent, Protect, Detect and Defend a company against unauthorized actors—internal and external. A CISO has to protect information within a company, establish rules of the road, employ governance and monitoring techniques, defend against attacks from inside and outside the company and allow only authorized individuals or systems to access a company's information infrastructure. A CISO and his or her team could be compared to the Department of Homeland Security but I think a comparison with prison guards is more appropriate. A CISO has to ensure that inmates behave in an orderly manner, there is sufficient segregation of high risk inmates from low risk inmates, only authorized visitors are allowed to enter and leave a facility, any communication between visitors and inmates is monitored and recorded for detection and analysis, only authorized goods exchange hands and so on and so forth. Of course, the reference to inmates is for systems, data and employees in an organization and the reference to visitors is for outside systems, data and employees of business partners as well as unauthorized parties such as hackers and social engineers. A CISO's job becomes significantly simpler if a company's systems, data and employees can be insulated from the outside world but we all know such companies simply do not exist. This makes absolute security impossible. Anyone who thinks that their company or data or systems are completely and absolutely secure is living in a fool's paradise. Security is all about building virtual layers upon layers and embedding multiple alarms. The goal of the these layers and alarms is an expectation that unauthorized actors would trip at least one alarm and abandon their quest or leave enough footprints that detection and legal prosecution become possible.

A company's supply chain extends far and wide and technology advances and creative usage keeps CISOs on their toes 24x7. One of the many important responsibilities of a CISO is to ensure that data is secure and preventive techniques are implemented to minimize risk if someone does manage to cross all the layers. One of the common techniques used is data encryption. Data encryption is a technique used to mask and create jumbled versions of data. Depending upon strength (as in robustness due to complexity) and type of *masking* that data can be reverse engineered to arrive at its original, readable state or not. If encryption cannot be reverse engineered then the only way to determine validity is to take unencrypted data, encrypt it and then compare it with the encrypted data in its stored state. If they look identical then you have a match. If not, try again. Why are we very briefly describing encryption? There's good reason. Determining characteristics and attributes of data such as the following need significant inputs from CTSO, CDO, CEO, Business Unit leaders, and CISO:

- Should data be encrypted?
- Which data (beyond the obvious) is sensitive to a company and/or its customers?

- Which data constitutes intellectual property or a closely guarded secret and so on and so forth?

Numerous tools and techniques exist that allow data and systems to be rendered highly secure. However, security comes at a significant cost in terms of financial cost as well as operating cost. Financial cost is obvious but there is a significant operational and perception cost as well. What I mean by that is the more secure data is made the more difficult it becomes to retrieve and share with customers or business partners. This tends to be in direct violation of everything business leaders, especially those in Marketing and Sales, demand. Business leaders operate under the assumption that information should be available readily and immediately to anyone and everyone who needs it or collecting information in not an entirely secure manner. Sometimes even at the expense of taking on undue risks, intentionally or unintentionally, to complete a sale or reach a target. This is further aided by a lack of awareness among consumers. Awareness about Information Security has increased over the years (thanks to efforts from governments, not-for-profit organizations, and business and technology professionals) but it is not yet where it needs to be. Companies, especially those in financial and healthcare industries, have realized that Information Security, while essential for all the right reasons, is also an essential component of a marketing pitch partly because it creates positive public perception and partly because regulated companies are required to follow applicable regulations/laws.

Business is hinged on balance sheets, profits and losses, and taking risks whereas Security is all about eliminating risk. Without risk there is no reward and risk translates to cost whether it is financial cost or opportunity and reputation cost, if it does not pay off. As a result, business leaders and CISOs should work hand in hand so business can move forward and grow. Business leaders should be educated by a CISO, via one on one education sessions, if required, and CISOs should understand core business operations so they can together help the company move forward. Without this partnership they tend to play *ping pong* with issues and in the end important issues either fall through the crack or take too long to resolve. Neither of which is an efficient or desirable outcome. Ensuring that such partnerships stand the test of time and allow businesses to remain nimble, run efficiently with the right risk mitigation plans is not easy but can be done. A simple way of creating this is through the same methodology discussed in the chapter on Requirements and Implementation. The Technology Interfaces Standards Template should serve the needs of performing preliminary evaluations and reaching conclusions on basic questions. Teams, very often, spend significant amounts of time repeatedly asking and answering the same questions. CISOs and CTSOs should publish information that educates and guides teams and making that information accessible from all the tools and processes used on a day-to-day basis. Many organizations create FAQs or compile documents that serve as reference guides but then make the monumental mistake of not making that information easily accessible and not keeping that information current. They might as well not have wasted time in compiling that information to begin with.

Chief Software Officer (CSO)

A CSO's primary responsibility should be to look at TAOs and ensure that the company benefits from cross pollination of knowledge, tools and people. CSOs should be well versed in designing and building business software that not only benefits business lines but also corporate and the technology department itself. Without increasing efficiencies of technology and other departments providing corporate support the overall company cannot achieve optimum efficiency. A CSO's job should be to scout processes in all departments in order to automate and deliver efficiencies. A CSO's position should be analogous to a quarterback reviewing the entire field of TAOs (and BAOs) and ensuring that they are all moving toward the end zone. A CSO should be someone who has cut his or her teeth in designing, architecting, developing, testing and, most importantly, successfully rolling out software–large and small. Employees, even those at the front lines, should have the faith and confidence that that CSO is looking out for them and that he/she knows their business, their needs, wants and pains. A CSO should spend sufficient time out in the field to watch, learn, and experience what various departments', employees' and customers' needs are. Without such a firsthand experience CSOs would have to always rely on someone else describing the problem, setting priorities and providing requirements.

Chief Technology Operations Officer (CTOO)

A CTOO's responsibility should be to develop awareness and monitor traffic flowing through a company's network. Drawing an analogy with a freeway, a CTOO should monitor if all drivers are following rules of the road, if there is congestion as a particular on-ramp or off-ramp, are there peaks and bottlenecks at certain times of the day or days of the week. Depending on the analysis a CTOO should add more lanes on the freeway in accordance with the business and corporate strategy. A CTOO should never wait for the drivers on that freeway to raise their hand and state that they need help. CTOOs in conjunction with CSOs and TAOs should ensure that all systems follow a standard set of automated reporting criteria. This consistency allows easier and standardized monitoring to allow consistent responses and quantifiable actions. CTOOs should be monitoring every entry point into and every exit point out of a company's network. CTOOs, in conjunction with CSOs and CISOs, should be responsible for orderly, efficient and monitored flow of information. Without understanding patterns and trends of information flow there will always be unanticipated problems resulting in revenue loss, reputation damage, and increased risk.

In summary, it suffices to say that IT leadership has to understand day-to-day challenges to effectively solve business challenges, tap into opportunities and avoid becoming custodians of data centers trying to maintain status quo. Companies spend 50-80% in maintenance/care and feeding of systems and only 20-50% in positive, net new change. This is a ridiculously low number in terms of effectiveness and productivity of IT organizations. Technology leaders should not be averse to taking risks or experimenting with new technologies. Otherwise, employees that see value or have an interest in using a new system, tool or technology will find a way to use them and then IT is stuck with either "maintaining" it or trying to exert their authority and figuring out ways to squash that creativity. There is no question that there have to be controls and checks but when IT does not solve real life problems a *black market* is bound to raise its head. If IT builds a reputation of solving problems and not giving people a run-around, circuitous processes, and poorly designed policies and procedures employees will be more than willing to approach IT with problems and expecting meaningful solutions, and accepting IT's feedback as one that is in their best interests.

8 I IS FOR INNOVATION

I WAS ONCE *COACHED* by one of my managers that to climb the corporate ladder and be successful in the company I should only focus on *high level stuff* and not on details. My knee-jerk reaction to that was, "How would I know what the high level stuff is if I do not know enough of the *low level* details?" That manager did not have an answer to my question but he reiterated that getting bogged down in details would prevent me from expanding my thinking and limit me from taking on additional responsibilities. If every manager started following that philosophy then who would delve into details and deal with the devil that lies in the details? What that manager said might have some truth to it when looked at in isolation. However, I squarely disagree with that philosophy. This philosophy implies that such managers can only speak with other managers and professionals that do not need to, do not care to or do not have the capacity to understand details. As a result, everyone will be comfortable within their four walls without wanting to know what's really happening in the real world. Successful managers and true leaders have to be as comfortable speaking with an analyst, a developer, a tester, a CIO, a CFO, a CEO, and Directors of the Board.

So, what *level* should managers operate at? It is, of course, not effective for everyone to be operating at the same level of strategic or tactical granularity but all managers have to have the ability to understand all levels of details. Based on their role in the organization different managers would have to focus on different levels of detail. However, organizations should take a hard look at managers who always are surrounded by an entourage *that knows details*. Such managers do not have command and control over the area they are expected to manage, do not understand the company's business, or do not have the capacity to learn. A big part of being a manager is the ability to connect the dots between Strategy and Tactics and being able to explain it to others so everyone is on the same proverbial page. When managers bring an entourage with them they essentially expect others in that entourage to understand what was discussed and what the next steps are. Have you seen and heard managers that ask their entourage, at the end of meetings, "you got all that and know what needs to be done, right?" Such managers are truly an overhead and eat into a company's productivity because someone else is carrying their dead weight.

Professionals who can straddle both strategic planning and tactical execution are a rare breed and organizations should groom or hire as many such individuals as possible. Such business technologists can truly help a company sustain profitability by being visionaries, taking calculated risks and knowing which solution has a qualitative ROI versus a quantitative ROI. I have come across many managers who are talking heads and know how to quote policies but have no idea how to administer them or determine which policies make sense and which ones do not. Overall, such talking heads stifle innovation and progress. Tell-tale signs of such managers are that they rarely have an answer to any question and their typical response starts with, "I need to check with my team…" Such managers, typically, know how to talk the talk but have no idea how to walk the walk. They either do not realize or realize but do not care how hollow their words sound.

IT professionals have to constantly reinvent themselves and be innovative in all aspects of their life. Innovation and reinvention do not automatically mean inventing new products. Innovation could, at times, be improvisation while being very aware of the pace and change happening all around. IT should be so aware and entrenched in the company's core business that IT, in conjunction with Sales and Marketing departments, should be able to propose, market, and "sell" products. IT can and should have a Sales and Innovation arm and that's how it can take strides toward becoming a profit center. Imagine the possibilities if IT were led by someone seasoned in technology but with a Sales and Marketing bend of mind or at least some experience in that arena? Just imagine the possibilities! Overnight, IT could become an innovator and not a cost center or be perceived as a drag on company resources. Of course, that enthusiasm would need to be balanced with rigor of disciplined processes and risk management but this is another way it can become a profit center. More often than not IT seems to be content in performing repeatable, mundane tasks and very comfortable in asking someone else to dictate what they should be doing. This is not how current day technology teams should perform and thrive.

Let's look at things another way. What is one of the key factors that interviewees are evaluated on? Is the candidate a self starter, takes initiative and works under minimal supervision? Some candidates say and actually are/do all those things whereas others just say the right things during the interview. Now compare that with the number of people in your team or company (including you) who truly do that and the number of supervisors that encourage or allow that to happen. It is very easy to talk the talk and use management lingo but when it comes to walking the walk it is a whole different ball game. It is as though people completely forget what they ask or answer during interviews, what they commit to doing and what they actually do on the job. Not everyone is capable of designing new products, services or processes but can companies identify those that have this capability? Can companies encourage and nourish such abilities? Of course, they can. So, why don't they? Companies, in this context, are nothing but people. People are grouped into teams. Teams have managers. Most managers are afraid of losing their reputation and territories if members of their teams are truly more smart, driven and capable than them. Moreover, managers do not necessarily understand what their role is. It is a very common perception that their job is to delegate, *manage* tasks, provide status reports, and, generally, maintain status quo or react to someone else's actions. This managerial behavior is very pervasive in IT and non-IT teams alike.

People are usually very good at thinking through issues from their perspective. Seasoned managers have full command and control over various scenarios when looking at them from their lens. However, that is never enough. True leaders are those who can put themselves in someone else's shoes and figure out what works, what does not, challenge their internal and external customers might face and so on and so forth. Moreover, leaders should not only challenge their own thinking process but also coach their teams to not get boxed in. You can review your email archives to understand the types of questions your customers and team members ask or the types of comments they make. This will help you understand if your analyses, assessments, and comments are sufficiently thorough. Your approach needs to be revised if your responses leave your audience wanting more information or with more questions than answers. If people ask you more questions or request more information that is an automatic extension of information you provided then your analysis was not sufficiently thorough. One should always try to provide comprehensive solutions and not just answers to questions asked. One should learn to read between the lines and try to understand the motivation behind the original question or information request. Only then can you be a solution provider and rise above being an answer clerk.

To truly be solution providers technology professionals have to learn to incorporate two attributes—Innovation and Prediction. While innovation is a critical attribute for the entire IT organization as a whole, it is also a much needed characteristic for individual growth—personal and professional. The following are critical questions that everyone should think about and answer.

- What do I contribute?
- What are my core strengths?
- What are my areas of improvement?

Make your mirror speak to you and try not to hide behind a veneer of callousness or wave them away with casual disdain that is typical when people do not want to face the hard questions. Most people who are talented and/or ambitious believe that the only path to grow and rise on the corporate ladder is by becoming a manager. This is mostly because, typical, IT organizations do not have career tracks that maximize a person's core strengths while providing growth opportunities. There's usually a singular track offered to technology professionals. Be a "worker bee" or get into "management." Technology leaders and teams have to get better at predicting needs and wants of employees as well as customers. In some ways, technology professionals have to know, within reason, what will be asked of them even before it is asked. However, not every manager can be expected to have this capability either based on their experience, capacity or due to other reasons. As a result, managers should be assigned various tactical roles that are rooted in relevant strategies. Managers can be classified as Resource Managers, Implementation Managers, Design and Development Managers, Maintenance Managers, Coordination Managers or Strategy Managers. One role is not necessarily better or more important than others but recognizing differences is very important. Without becoming aware of such differences, a manager could be forced to play all these distinct roles resulting in delayed and expensive execution. Football (American Football not Soccer) has distinct teams for each segment of the game— Offense, Defense and Special Teams. Each team and individuals within that team have a distinct role that they have to play to maximize a team's chance of success. Unlike Football, IT organizations assume that all managers are alike and that one manager can easily perform the role of any other manager. Management has many different facets and just giving someone the title of a manager does not automatically make that person a manager make. Management is comprised of team management, customer management, budget management, quality management, service management, morale management, growth management, and cost management to name just a few. Just like Football, IT should create well defined teams with specific contribution expectations. It would be absolutely impractical and uneconomical to assign one manager for each area of management. However, expecting every manager to do a fantastic job in every area places an undue burden on individuals and most managers become mediocre because they simply get pulled in multiple directions at once and their focus gets so diluted that they eventually get disillusioned and just keep doing more of the same or leave hoping for greener pastures, which may or may not exist. If you ask IT managers and non-managers what their responsibilities are or what their contributions were the most likely answer will be that they completed projects A, B and C or that they managed a team. Irrespective of the specific answer, it would indicate that they operated in a very tactical manner and did not realize how their contributions helped the company grow or solve real world situations.

This type of tactical thinking is partly due to the fact that, in many companies, managers and descriptions of job functions do not change unless there is a significant event in the company forcing an organizational change. As a result, there rarely is a change in people's responsibilities. Just as Bob and Bob, in Office Space, are engaged to cut the fat out of Initech, companies should undertake annual reviews of all managers to determine if they continue to be good fits. I am not suggesting that such reviews be used to threaten managers or keep them on their toes but to understand if their skills have changed and if their role continues to align with the company's needs. If managers cannot coherently answer what they do then that should be considered a red flag. *I manage people* should not be an acceptable answer. Managers have to be able to explain to their managers what they believe their contributions are and team members should be able to appreciate what their manager does for them and how he/she contributes to the company's bottom line.

The point I am trying to make is that companies have to take a hard look at their IT organizations and IT has to get away from thinking about projects and innovate from the ground up. Technology teams have to justify their existence, methods of thinking, communication style, and contributions to the company's bottom line. In a nut shell, IT has to understand and explain its unique value proposition (UVP) without which IT might as well be a call center that waits for customers to call so they can give them canned answers. Companies and IT have to determine what kind of innovation is important and relevant to their situation and ask themselves one simple question. What kind of innovation do we need this month, quarter, or year and how are we going to achieve it? There's always room for innovation and never a dearth of improvements needed or problems to solve. It is likely there exists only a dearth of commitment to accept existence of problems and a personal and corporate commitment to innovate.

Part III

THINGS PEOPLE DON'T THINK ABOUT

9 PERCEPTION IS REALITY

When you think of an executive chef what image comes to mind? How about an airline pilot? The image these professions project is one of confidence, complete control over their craft and above all excellent communicators. This image and attitude makes their customers feel at ease and as though they are being provided the highest level of service with nothing but their satisfaction in mind. Now think of a typical knowledge worker in the IT industry whether they work directly for your company or another company. Think of programmers, managers, testers, analysts, help desk or service technicians, call center agents, etc. What image comes to mind now? Depending upon the organizations you have worked in and whether you yourself are one of these IT professionals or not the image that comes to your mind will vary but only slightly.

Most IT professionals are notoriously poor communicators and do a poor job in presenting their ideas and thoughts, however brilliant they might be. That's where non-IT managers and consultants add *value*. They take sound bites, re-order content, add certain buzz words, top off presentations with pleasing colors and pictures and present it like it was their brain child. People consuming that information are none the wiser or do not care; sometimes rightfully so. There is nothing wrong with providing a service when there's a demand. However, the point simply is that people doing the leg work and who these ideas really belong to should be presenting them and getting their due credit. Otherwise, the knowledge gap between the idea owner and the presenter could result in unintended consequences. Such presenters, or information brokers, typically, do not spend the time and are ill-equipped to understand the genesis of an idea, fundamental principles, benefits, and constraints. As a result, when they are selling ideas they tend to over-commit what can be done, whether something should be done or undersell the power of that idea. That, in turn, results in customers being promised services that are very difficult to render or being sold a bag of trinkets.

How often have you been in situations where your IT problem or question is answered incorrectly over and over till you get to that one person sitting in an obscure corner of the organization? Once you find that person questions get answered, problems get solved in almost no time. What should have taken a matter of minutes or hours takes days, weeks or months because organizations hire and build layers and layers of management. Why? Because they believe that that person in that obscure corner needs people to manage him/her and to be a liaison with the world at large. These liaisons tend to focus on being friendly with their customers because that's their primary survival skill. However, sooner or later, mostly much later, the true value of services these liaisons provide comes to light and organizations then take the radical step of replacing them... with another liaison! This kind of locked and boxed thinking does nothing but diminish the value that customers receive, increases operational costs and leads to loss of true talent. How, you ask? The many layers of people between the true source and beneficiaries of such information are bound to cause loss of information in translation. Moreover, the number of layers will be directly proportional to the amount of loss. This is just a natural phenomenon similar to loss of energy that is directly proportional to the distance it has to travel before it reaches its destination. It increases operational costs because these layers, typically, command higher pay and job titles than people in those obscure corners. It results in loss of talent because these layers block financial and organizational progress of those that are true generators of ideas, creators of knowledge, and disseminators of value and information.

Instead of building and re-building these layers organizations could very easily invest a fraction of their operational costs in coaching and training those buried under the layers and give them some sunlight to trigger and catalyze their growth. I guarantee that these organizations will deliver tremendous value to their customers, lower costs and increase revenues. Organizations claim that people and human resources are their greatest asset but that is mostly management-speak and organizations do not realize the true value of that tremendous talent pool. One reason they do not realize or tap that value is because leaders in ivory towers and with C-titles believe that their job is to stay at a "high level" and not have to stoop to low levels to delve into gory details. My question to all those that   believe that details should be left to mid-level managers and lower level employees is very simple. If you do not understand any details then how do you believe that the higher level of information you have been provided is accurate and how do you communicate that to your bosses and customers? If the answer is that you provide information as it was provided to you then what value are you adding by being a middle-man and a broker? It is common sense that decomposing higher level or summary information is difficult or sometimes impossible but summarizing from a more granular level is easier and logical. People that are true liaisons add value by thinking about things in a different way or providing a different perspective, analyzing situations from angles that have not been previously considered but if the purpose is to simply "forward" ideas, emails, and reports then such people are organizational fat that should be trimmed.

My advice to leaders in IT is to try and understand the true value of each and every employee in your organization, identify their strengths, weaknesses, and flaws and then determine how you can maximize their strengths and eliminate their flaws by training or coaching them. You will get long term loyalty by giving the right employees the right opportunities and maximize productivity like you have never seen before. The other big benefit of this approach is that you will receive accurate and timely information and that translates to speed of execution, effective communication, higher trust and reliability with your customers. All these can only improve the bottom line.

On the other hand, my advice to IT professionals is to look in their professional mirrors to determine why information brokers are needed. If you are one of those professionals that like to program all day, every day and would rather debug code than speak with people then that is your choice and there's absolutely nothing wrong with that. However, if you are one of those people that like to add or improve your softer skills then you have to take a very hard look at yourself. Be very objective, record your voice on the phone, leave yourself voicemails, record your body language using a video camera and then perform a self-evaluation. Watch how reporters, hosts of prominent, reputable TV and radio programs present themselves and then compare your presence and presentation skills to theirs. The difference in most cases will be immediately visible. Those reporters and hosts were not born that way. They practiced their craft, their voice, their delivery, and their facial expressions and so on and so forth. They mastered such skills with time and practice and so can you. Pick one skill at a time and focus like a hawk on its improvement. Be very objective and brutally honest with yourself. Pick a partner who also needs improvement and is committed to improve. Such a partner can give you objective feedback and can also receive objective feedback from you. Once you figure out what's lacking and what you need to improve start practicing your improvements on a daily basis. Remember, time and practice are what allow people to master such skills. So, don't be shy or hesitant. People will notice the changes you are trying to make and will be pleasantly surprised. A lot of them will even commend and encourage you. Lastly, if you think you need professional help in improving your skills then do seek that help. Do not short change yourself but invest in yourself because it will be well worth the investment.

When I was trying to improve my public speaking skills I learned very early on that the only way to test myself was in public. One can practice in the comfort of one's home but you will never know if you have improved and how much improvement there is if you don't test yourself in public. When I was just starting out as a freshman in college I went through a phase of self-doubt and wondered if I had what it took to get through the bachelor's program. So, I approached a friend of mine who was a senior in the college. Admittance to good universities was determined by one's rank in a state-wide common entrance exam called EAMCET. Those that were in the top 1,500 out of 100,000+ students who attempted that exam gained admission to the university I graduated from. My friend asked, "What was your EAMCET rank?" I said, "749." He said, "Do you think every single person in your class is smarter than you and more talented that you?" He continued, "If you think you are not as smart, if not smarter, than most other students then you should drop out immediately. Otherwise, have faith in yourself. Accept your shortcomings, work on them and improve yourself. Similarly, know your strengths and talents and let them shine through." Those words continue to have significance in my life to this day.

Each and every person has strengths and weaknesses. Perform your personalized, individualized SWOT analysis to learn and improve in order to achieve your goals whatever they may be. Most of us say what we want to do and how we are not able to achieve our goals. The simple fact remains that people do not focus on figuring out how to get to one's goals. No self-help book or guru can help you achieve your goals by practicing their prescribed techniques. It is easier said than done but all that's necessary is raw courage, a no-regrets-and-no-shame approach to introspection, identifying sacrifices that need to be made—however small they may be—and where extra effort needs to be made. Finally, one has to be absolutely and brutally honest with oneself, and then take huge swings for the fences.

10 TECH CULTURE

When my wife and I were expecting our child we became avid readers of books about babies. We read books about prenatal care, infants, toddlers, and everything in between and beyond. Every book had one common theme. Every author and expert made it abundantly clear that the atmosphere at home directly affects a baby's mood. They said parents should be cheerful and loving and the atmosphere should be bright and airy. These factors directly affect how happy, responsive and interactive a baby will be. Conversely, parents who do not interact well with each other or the baby and live in dimly lit houses risk causing depression even among babies. What they mean is that the culture at home tells even infants and toddlers how they should react. The culture of an organization and its effect on employees is no different.

Organizational culture is directly determined by the most senior management—in some ways, legal guardians and caretakers of an organization. Most organizations have a top down effect in terms of culture, ethics, fairness, principles, and customer service. The way executive management treats its employees is how employees treat their peers and customers. Irrespective of the size of the team, every person in the team learns, for better or for worse, from the person above them. Being an organizational exception to this tendency can be very fortunate in some cases. People like these can bring about well meaning and, often, necessary changes. Companies that have been in business for a long time typically have a number of managers that rise through the ranks to a culture that has been the norm for several years and decades. Organizations, and ultimately their customers, can benefit immensely by creating a refreshing change in the way processes are run, motivating employees and piquing their interest by challenging them. But, the culture of change has to be inherent in management. They have to be able to recognize the positive effects of change and be able to pick the right people to drive change. Whether people like change or not, they thrive in it and this one simple thing escapes most managers. Managers, especially those in upper management, have to show that they themselves can change. Otherwise, it becomes another case of talking the talk and not walking the walk. One of my friends used to work one Saturday every month. I had known his routine for a long time and one day I asked him what he did on a Saturday that he could not accomplish on Friday or the following Monday. He said that he had to generate a dashboard report a certain way and that took almost an entire day. Upon further discussion we both realized that there was another, more efficient way to generate and clearer way to present the information. His response was that he had worked with Jack (not the manager's real name), the head of his department, for over 7 years and that that's just the way he liked it. There was no way he would accept it any other way and that he was set in his ways. I was not really surprised to hear that because humans are creatures of habit. However, it is no surprise that even leaders who are supposed to motivate their departments, lead from the front and show people the way fall victim to such inefficiencies and make others inefficient as well.

Not recognizing the need for change can have an adverse impact on everyone involved. It can demoralize, kill enthusiasm, and make people fall into an "acceptance" mode. This results in high employee turnover and performance below par. This, often, happens when teams have been managed for several years by the same managers. This does not necessarily mean that experienced managers should be replaced by people with newer, different experience levels. On the contrary, managers with experience are an asset to organizations. They accumulate a wealth of knowledge over a long period of time and their direction in times of crises is crucial. However, experience should be upgraded and complemented with skills that are core to their functional role in the organization. Management styles have changed over the years and research has identified new techniques that help leverage strengths, resolve conflicts, and celebrate diversity. Classes that teach social styles and technologies, of course, continue to evolve on a daily basis. All these are essential skills in these ever-changing times and multicultural, global workplaces. Twenty-five years ago employees were supposed to be at work at the same time every day, stay eight hours and leave at the same time every day. The longer they stayed the better they were in the eyes of supervisors. There were strict dress codes, men and women were instructed to be a certain way and there were mostly men in management positions. Twenty-five years later, everything has changed—well, almost everything. Employees can telecommute, have flexible hours, dress codes are more casual in several places, women are corporate leaders and several culturally diverse people work together. All are changes for the better and long overdue. However, in all those years the definition of a manager or how managers are identified, evaluated, and promoted has not changed. This unfortunate lack of progress in management skills and outlook is disastrous for teams and companies. Management doesn't recognize the need for skills to be upgraded and complacency sets in among employees. They tend not to go above and beyond their job, they do not apply their skills to their full potential and many leave resulting in significant costs for companies. This is a culture that can rot the very roots of a healthy company.

Almost all management books and seasoned managers make a standard comment that employees are the most valuable assets (MVAs) of a company. Employees truly are and they should be treated as such. The reality, unfortunately, is very different in most organizations. However, some companies do live up to this statement and have successfully retained their employees for several years, if not decades. The common sense approach of (smart) employee retention not only reduces expenses but allows companies to maintain a deep knowledgebase, a fiercely loyal and trustworthy group of employees. When employees are treated with respect and fairness they will invariably treat their customers with the same degree of respect and fairness. Most organizations, though, believe that the first step to cutting costs is to reduce employees, cut payroll and decline promotions to the most deserving employees. Although in some cases this is inevitable, in a lot of cases it can be avoided. Understandably, payroll represents one of the biggest costs in most organizations but when that hand is overplayed either wrong people get laid off or wrong positions get eliminated. In most cases, decision makers and line managers do not have the ability to recognize talent or understand who is capable of what or what their experience has been within and outside the company irrespective of what their current job responsibilities are. Managers should keep a profile of each of their team members to avoid making bad decisions. This profile should contain their technology skill set, experience in various businesses (past and present), other skills/certifications such as project management, and technology certifications. This profile should provide managers and peers alike a view into what each person has achieved at the company, what skills they have that are being utilized and those that can be utilized, experience that can be leveraged as a backup to someone else's primary skills and so on and so forth. Almost always, creation of a new position results in internal and external job postings on job boards, internal company sites, or searches through placement firms. How many managers and, more importantly, human resources (HR) departments really know about skills available within a company? Hiring managers, HR recruiters and HR managers have to stop thinking like their predecessors in the 80s and 90s used to think. Where is the innovation in true people management and resource management?

Ask HR and hiring managers what their personal MVAs are and how they take care of them. People have cars, TVs, jewelry, wine collections and so on and so forth as their personal MVAs. Now ask them how they care for them. Cars are cleaned and buffed every week, wine is stored at just the right temperature in a nice wine cellar or refrigerator, and TVs and jewelry are cleaned, handled with white gloves. How are employees–corporate MVAs–handled? How are these MVAs cared for? Here's how. They are told what to do, how to do, and are very clearly told who's the boss. Managers like to exert their authority mostly for the wrong reasons. Why do line managers do that and why do executive managers–those with C-level titles–allow that to permeate throughout their organizations? Is it because they are blissfully unaware or aware but do not care as long as they do not have to deal with problems resulting from such poor handling of their MVAs? So, is saying employees are their MVAs just something fashionable to say, just management jargon or do managers really believe it and understand what they are saying and how to prove that they are not just empty words?

Employees and staff should be treated with a great deal of care, respect and responsibility. When I was in ninth grade there were two kids in my class who were considered troublesome by most teachers in our school. Where I went to high school, in India, we used to have a Class Teacher. Class Teachers were responsible for the overall well being of kids in classes assigned to them. Our Class Teacher was relatively new to the school and not only was he unable to handle those kids but other teachers had also complained to him about those kids. Not knowing how to deal with the situation he approached another teacher who was much more experienced than him. He asked her, Sir Charles, how he could control and discipline them. We used to call her Sir Charles after Sir Charles Baskerville from The Hound of Baskervilles–the classic Sherlock Holmes story. She used to teach us English and she used to teach us fabulously well. Her name was Ms. Charles and we affectionately used to call her Sir Charles–of course, behind her back. Coming back to the point, she told our Class Teacher that he should assign those troublesome kids real responsibility. She said make them class monitors or assign another task where they become responsible for the behavior of all the other kids in the class. As a result, they would have to shape up and be on their best behavior in order to become examples for other kids. Lo and behold they were no longer mischief makers. They were mischief controllers and became responsible students almost overnight.

I have remembered that incident ever since and thought that if I ever saw a miracle then that was it. The reason I am relaying that story is that employees are the same way. Instead of assigning them to work on projects, give them the responsibility to identify problems and solutions and you will see real results. You won't need as many managers who tell them what to do and how to do it. In this context, I am referring to managers as those that just attend meetings and forward emails and do not contribute anything of real consequence. We all have seen and had such managers on more than one occasion. So, how do you assign responsibility? It begins with hiring the right people.

To begin with, corporations must have strict and carefully screened hiring policies so only people that can contribute to the overall well-being of the organization are hired. Companies have to completely overhaul the way recruiting is organized because in most cases, the recruiting team is directly responsible for the kind of people that get hired. Most corporate recruiters have no incentive in finding the best candidates. They frustrate the hiring manager and in effect crippling the whole process. On the other hand, hiring managers do not necessarily take the time to provide recruiters with the information they need to be efficient. The problem, in a nutshell, is that if people in the value chain are simply brokers and information forwarders then they will never be able to truly provide valuable contributions and others will not help you perform your job any better. So, it becomes a vicious circle wherein everyone has many gripes and complaints about everyone else but no one does anything or understands what needs to be done to solve the overarching problem. Everyone in the chain has to have some skin in the game Imagine what would happen if a recruiter's performance were directly tied to performance of employees he/she helped hire. Let's take this to a completely different level. What if there were no hiring managers? Who truly needs bosses? Would employees and staff be more motivated and if they knew that they did not need to worry about someone watching over their shoulders and telling them what they are doing wrong or why they need to do things differently. In most cases, bosses are considered to be just delegators of tasks, projects, problems and not true contributors. I have seen many, many managers who cannot take part in real discussions about solutions and problems without bringing an entourage with them. For everything their answer is, "Let me check with my team" or "Let me speak to the expert in my team." So, what exactly do these managers do or contribute? Many people have said that they barely speak with their bosses and when they do they tell him or her anything they deem appropriate and the boss would be none the wiser. I strongly believe that most knowledge workers either know their job very well or can be informed and given the freedom to contribute in the best way possible as long as everyone understands the expected outcome and the strategic direction. The positions of managers should be converted into information aggregators and integrators of individual components. They have to play a real role, carry their own weight and not just rely on everyone else in the team doing the work. Managers should provide guidance, mentoring, applying lessons they learned the hard way so their team members do not repeat their mistakes.

People are usually very good at thinking through issues from their perspective. They have full command and control over various scenarios, pros and cons, etc. when looking at the world from their vantage point. However, it is very difficult for people to put themselves in someone else's shoes and determine what works, what does not, what challenges others might face, and so on and so forth. Such tendencies can be clearly noticed in salespeople, corporate strategists and many, but not all, C-level executives. People should almost always put their thoughts on paper, draw flowcharts, create process flow diagrams and put themselves through the process that they are creating for others to follow. Process creators should always apply the process to themselves and role play as a user of that process instead of looking at it only from the eyes of an administrator of that process. People miss the point that designers and administrators of a process use those processes only rarely, if at all. Just because a person designs it on paper and throws it over the wall for others to use it does not mean the process is perfect or does not need revisions. It is one thing to design it and completely another to actually use it day after day in the real world. People have to know that unless they eat their own cooking they cannot assume that they are ready to be Top Chef even if one or two people with no taste buds praise your culinary skills.

People have to understand that designing a process or policy at a "high level" and not getting down into gory details is not necessarily an attribute. If every salesperson, manager, delegator, administrator tried to follow his or her own sales pitch, delegation policies, management practices they would find so many flaws and inefficiencies that they would wonder how people on the receiving end had been dealing with that process year after year. That's the beauty of being a manager. It tends to be a one-way street of information flow. Managers have to learn this habit of knowing that every word that comes out of their mouths, every email that leaves their keyboard will be memorialized and that people you never thought or wanted to see your emails will, sooner or later, see them. They have to listen to what they say, read what they write and not adopt a fire and forget attitude. My all time favorite fire-and-forget line is, "I accept responsibility and we will ensure it does not happen again." Most people that utter such statements do not have the faintest idea what they are accepting responsibility for and what, if anything, they plan to do or what their individual ideas are to prevent the types of mistakes in question.

Technology organizations have to create a culture wherein every person carries his or her own weight and should be able to clearly state what their individual contribution was and how that contribution helped the team and, ultimately, the company. Managers should not be able to always hide behind, "my team..." Without such a culture being adopted right from the top, it is difficult and unfair to expect the rest of the organization to be transparent, accountable, and a self-motivating force.

11 POLICIES AND PROCEDURES

IT IS DIFFERENT FROM most other departments/groups in a company. Most departments are guided by policies and procedures that clearly define what, how and why employees are supposed to do what they need to do in various situations. If people follow those policies and procedures then the outcome is almost guaranteed. In other words, those policies and procedures are a playbook that employees have to follow. On the other hand, IT is quite different. Policies and procedures go only so far. Beyond that it is up to people to ensure things are done right. Audits can be performed to ensure accounting is accurate, loans are underwritten correctly, medical compounds are mixed correctly, car parts are manufactured correctly, and so on and so forth but IT tasks—security controls, software development, hardware configuration, etc.—are usually not audited in as much detail as they should. Moreover, writing policies and procedures for IT is like trying to write policies and procedures for composing emails. A large number of ways in which a technical goal can be achieved make the focus be on the end goal and certain milestones along the way but not necessarily on the pathways to the goal. As a result, it is very possible for employees to take, say, twenty steps when they could have taken just eight. Add to that the fact that policies and procedures are almost always created and maintained in the form of documents and spreadsheets that are stowed in some arcane place that is hard to find and even harder to determine their currency.

The obvious question that comes to mind is, "Why doesn't IT change instead of continuing to think like it is operating in the 1980s?"

IT organizations have devolved into organizations that have become slaves to their policies and procedures instead of making those policies and procedures work for the benefit of the company. This results in IT personnel saying No more often than saying Yes to solving problems or providing solutions. This attitude, in turn, results in business managers, customer service personnel and customers alike sending a vibe that IT is a black hole with no hope of getting timely resolution. As important as policies and procedures are technology leaders have to identify which policies are truly important in reducing risk, increasing efficiency, etc. and which ones are simply creating barriers. A very commonly used phrase in books and speeches on leadership is that leaders should eliminate barriers and empower their teams to foster innovation and creativity. However, leaders who read such books, listen to such speeches or dispense these words most likely have no idea which policies and procedures their teams follow and which barriers are preventing innovation and growth. As a result, it is unlikely that they would know which barriers need to be broken down. On the flip side, employees continue to suffer and most of them do not bring up their problems and if they do they communicate them in a way that cannot be quantified (in management terms) or their communication gets lost on one of the rungs of the corporate ladder. This stalemate resulting from lack on information on one side and lack of effective communication on the other side prevent real change from occurring and entrenches companies in—you guessed it—mediocrity.

Just as we take our cars for regular maintenance tune-ups IT managers need to frequently evaluate how efficiently the IT organization is run and such reviews will give it the needed impetus to suggest efficiency improvement in the rest of the company. If IT cannot show that it improved its own house what right and *street cred* would it have to suggest efficiency improvement in other areas of the company? After all, charity begins at home. IT has to start with reviewing every policy and procedure and asking itself if a policy is a must-have or a nice-to-have, what purpose is it serving and, more importantly, whether it is still relevant.

Here's an example of how IT has to evaluate its operations and think creatively. Many companies have *going green* as a corporate initiative but they stop short of truly digging deep to identify all possible ways that create waste, increase carbon footprint or contribute to the planet's problems. Beyond the obvious tactics of sending company-wide emails encouraging employee participation and consideration, and adding tag lines to email signatures green initiatives are different for different departments. While each department has to focus on its specific contributions, most departments have policies and procedures in formats that are printer-friendly. Documents, spreadsheets, presentations, flow charts, etc. are all in printer-friendly formats that tempt people to click the Print button. A significant effort is spent on thinking about ways to prevent printing with thoughts ranging from monitoring who is printing what/how much to making access to printers difficult to eliminating printers altogether. However, in my opinion, that is an ineffective way of approaching the problem. In this specific case, companies should think about eliminating the need for printing by altering the format. In this day and age of iTunes and You Tube why don't companies think about utilizing audio and visual formats for policies, procedures and other such material. Most of us are used to consuming information in those formats, anyway, and our brains are probably tuned to absorb information faster when presented in those formats. How many of us have tried to print material available on You Tube or iTunes?

Accurately defining problems and creatively approaching solutions are highly underrated skills that companies should actively seek out and try to identify within the ranks. Lack of such skills or not recognizing existence/importance of such skills makes CEOs, non-IT staff and customers alike believe that IT lives in an alternate reality wherein customer needs and revenue pressures have little bearing on IT's modus operandi. IT is very good at creating plans and talking a big game but, often, short on execution or providing cost-effective, innovative solutions whether they are to further green initiatives or to bring Policies and Procedures in to the digital age. IT has to think about making technology work for IT.

12 MANAGEMENT TALKING POINTS

As PEOPLE *GROW* IN organizations and climb the corporate ladder some people grow smarter and wiser while others climb due to someone else's good graces. Irrespective of the growth path and wisdom of managers, many managers equip themselves with, what I call, MTPs–Management Talking Points. These eight (8) questions or comments are the most commonly used phrases by managers. Those aspiring to climb to upper management positions or interacting with upper management would be well advised to become familiar and be prepared to answer these questions in all situations. While many of these are relevant questions/comments, repeating this same set of questions in every situation is what sets true managers/leaders apart from those who are managers by virtue of just their titles.

1. What can we do to get it done sooner? Followed by... Can we throw more bodies at the problem?

2. What will a solution cost? Followed by... Are there less expensive alternatives?

3. I don't understand why this is still not fixed! It is terrible that it takes this long to fix this problem. We need all hands on deck!

4. I take full responsibility for this problem. We will do a full post mortem and will do whatever it takes to prevent a recurrence.

5. Thank you. Great job. Your hard work is very much appreciated.

6. Give me a status update. Followed by... What can we do to get it done sooner?

7. I want you to lead this effort and get back to me with a proposal.

8. Please work with Jane Q Analyst and John Q Programmer to get this done (with the implied message being... don't come back to me with any problems, just tell me it is done and that's all I care about, and I can then claim credit in front of others).

For many managers these questions/comments become Standard Operating Procedure (SOP) along with attending meeting after meeting after meeting and communicating mostly via emails. Someone once asked me what I did at the company and I said, "I attend meetings and forward emails." The person asking me that question had a surprised look partly because he did not expect that answer and partly because he must have thought I was out of my mind to state that in a room full of managers. I was, obviously, not worried about stating it because I knew what my individual and team contributions were. However, how many managers can look at themselves and acknowledge, even if only to themselves, that that's all they do? Managers spend an inordinate amount of time on status reports and conveying status back and forth weekly, monthly, and quarterly. In spite of spending so much time on status reports, a large number of managers are unaware of what the core problem is, what the solution is, why it is the right solution, and, more importantly, how it fits in the overall Strategy. More often than not that information is known to only a select few individuals and the rest of the people are just peddling status reports. Even trained pigeons can send around status reports and delegate work. The real challenge and value of a professional or a manager is in identifying and implementing long term solutions that eliminate the root of a problem or provide a solution to an opportunity before it become a necessity or a problem. If you are one of these managers who has an MTP approach to work then you should ask yourself, "Is this the only thing I do or do I have an expertise in one or more areas?" If you do have that expertise then ask yourself, "how has the company directly benefited from my expertise?" Delegating work or finding the right person to do a task is not necessarily a skill that should be considered one's area of expertise. It is surprising that a system or a mobile app has not already been developed that automates tasks of delegation. How would you respond if *Bob and Bob* (from Office Space) asked you, "What is it that you do around here?"

13 IT AUDIT 101

I AM SURE EVERY IT professional absolutely *loves* the sound of A-U-D-I-T. Whether employees, outside consultants, or others perform the audit, they all look for similar aspects–procedures and controls. Heed advice from someone who has learned it the hard way. When implementing a solution, engineering a process, or solving a problem always take into account factors that auditors look for. Think of this very short chapter as IT Audit 101. Answer the following questions as part of your system architecture and design and your Audit conversations will be easier, quicker and less stressful.

- Who approves authentication and authorization requests from users? Is that policy/procedure documented?
- Is there a policy for type, strength of authentication, and frequency of change? Is that policy/procedure documented?
- What steps are taken to ensure that those with approval privileges do not abuse them? This question comes up especially when a single person has the approval authority. Is there a report or audit performed at a pre-determined frequency? Is that policy/procedure documented?
- How is persisted data secured if that data is Non Public Information (NPI) or Personal Identifiable Information

(PII)? Is that policy/procedure documented and can it be verified using real life examples?

- Is data at rest and in transit secure and encrypted? Is that policy/procedure documented?
- Can system access be granted on a modular basis? In other words, can access be granted to only those parts of a system that a user needs rather than the entire system?
- If data is exchanged between two systems what controls exist to ensure that data did get transmitted as intended? How are such data transfers verified and who is responsible for tracking, communication and resolution when things go wrong? How is corrective action implemented? Is that policy/procedure documented?
- Is there a data archival and purge policy? Is that policy/procedure documented?
- If a system renders decisions based on automated rules/calculations how are those validated to ensure anomalies do not exist in those rules? Is that policy/procedure documented?
- If a system "refers" a decision for manual approval/override who makes those decisions, how are they recorded, where are they stored and who has the authority to override? Is that policy/procedure documented?
- How are processes–ranging from manual procedures to hardware builds to anti-virus engine/patch updates to Operating System service packs–monitored, communicated, approved, implemented, and verified? Are those policies/procedures documented?
- What are the procedures for deploying system changes to the live server environment? Is that policy/procedure documented?
- Is there separation of infrastructure, data, and personnel responsibilities between non-production and production/live systems?

Does this mean these are the only factors considered by IT auditors? Of course, not! Are these factors considered by most auditors most of the time? Definitely! It will be extremely beneficial for all IT professionals to think along these lines and be prepared with "controls," "procedures," and "protocols" to ensure that they plan ahead. That way, less time would be lost in retrofitting systems/processes and, thus, increasing time available for constructive, problem-solving, opportunity-grabbing solutions.

Early on in my career I used to think that such audits were a big waste of everyone's time but over time importance of such audits and verifications became clear to me. However, almost all auditors fail to look into certain areas or they spend so much time on the standard questions (listed earlier in this chapter) that they do not spend time on identifying real issues that lurk under the surface. Auditors need to be able to connect the dots from concept to roll out. Only then can they truly audit the entire process and the entire system. Auditing individual systems or processes in isolation does not help identify real problems, many of which exist at *process boundaries* or *process hand offs*. Often times, auditors lack the necessary know-how and background in their area of expertise, which makes them ill-equipped to truly dig deep and identify all potential pitfalls. Just as some of the major companies hire *white hats* to determine vulnerabilities in technology systems companies should hire individuals with programming, information security, hardware configuration, and such core technical skills. Hiring individuals with certifications and previous job titles does not automatically make a person eligible for such key positions. Technical knowledge, experience and certification should be considered when hiring for such positions. Having said that, how many IT professionals would say that their dream job is to become an IT Auditor? This is a quintessential example of how IT has to create a new image of itself and market it. I, personally, think an IT auditor's job is tremendously challenging and exciting when approached the right way. However, trying to follow best practices laid out decades ago and not inviting professionals to review and critique their approach is definitely not one of them.

All companies should ensure that their Audit groups publish white papers or guidance documents that provide an insight into what auditors would be looking for, what definitely not to do, which areas to tread lightly, and, more importantly, their capabilities and their thought process. When I had requested such a document from one of my ex-colleagues she stated that it amounted to conflict of interest and they could not publish such a document lest IT professionals used it as a guidance document and boxed in the Audit group. She said auditors do not like to lay down any rules or even guidelines because they prefer to be able to critique and make recommendations on a case by case basis without being held to any rules or guidelines. What that conversation told me was that auditors do not like to set their *requirements* for fear of being *audited* but like to critique everything. Even the Federal Open Market Committee (FOMC) of the Federal Reserve System publishes meeting notes and guidance but Auditors do not like to publish their guidance. That makes absolutely no sense whatsoever... to me. I am sure professionals smarter and more experienced than me might have another world view but that's where I stand. Like it or not.

Auditors have a very important role in maintaining safety and soundness of a company and, by extrapolation, the health of an industry and society at large, especially when you consider the financial services and healthcare industries. You can think of them as the cops of the systems and process world. Even the Los Angeles Police Department (LAPD) publishes a manual that lists their policies and procedures. So, why should IT cops be exempt?

14 TRUST AND PROFESSIONAL COURTESY

Trust is a very thin thread that can be easily broken but is very hard to weave and maintain. People have to work very hard each and every day on creating and maintaining a sense of trust and strong professional bond without which it is difficult to conduct business in good faith and have open lines of communication. Different people have different communication styles, different levels of comfort in public speaking and different levels of confidence in their friends and coworkers. Spoken words, written words, unsaid words and body language contribute to one's image and individuality. You should not try to be someone you are not but you should definitely be aware of what you are saying/ writing and how you are saying/writing it. People will perceive you to be a trustworthy character and a true partner or someone who would stab them in their back at the first opportunity. As discussed in previous chapters, most people are not born with polished "corporate skills" but one can learn by teaching oneself or seeking professional help. Writing emails to yourself, leaving voicemails to yourself, and recording your public speaking and presentation performances are some of the simple things you can do to measure yourself. Then put yourself in your coworkers', customers', and managers' shoes to objectively critique yourself. Without such a critique you could go through your entire life without having a chance to truly rise and improve your skills all the while wondering why you don't get *ahead*.

One of my former co-workers would talk negatively about everyone behind their backs and it always made me wonder what he said behind my back. Needless to say, most people never trusted him with anything and kept him at an arm's length. Trust is a very powerful and delicate element in day to day professional interactions so much so that people will make time for you or not spare even a moment even if they are twiddling their thumbs all day long.

Professional courtesy is another element that people do not give due importance to. People assume that others know exactly what they know, what is important to them is also important to others, and what is not their responsibility, in their opinion, is automatically someone else's responsibility. More importantly, it is not their responsibility and that is all that matters. The attitude to think about and respond to situations in a courteous manner that respects others' time and perspective seems to be seriously lacking in many *professionals*.

I get cold calls and emails everyday with arrogant statements about saving the company a significant amount of money, improving efficiency, reducing employee turnover and so on and so forth. Usually, those callers and e-mailers do not even know the right way to spell the name of the company but they talk with such bravado that you'd think they were promising to solve world hunger. The other classic pitch I get daily is, "… we have been doing SOA for 20 years and are pioneers in that field." My response some times to the callers and sometimes in my head is, "Really? Are you so behind the times that you haven't even caught on to the latest buzz words?" Ease of finding information through social media, professional networks, and mailing lists have made communication, rather spamming, very easy. As in most cases, people tend to lose respect and do not appreciate the value of things that come easily. Such things tend to get abused. I guess, that is the price we pay for the many true benefits of a connected, hyperlinked world.

Say, all these cold calls and emails do result in customer acquisition and gaining someone's business. How much focus is then placed on ensuring those customers' satisfaction and providing services exactly as committed? The focus and energy always seems to be spent more on new customer acquisition and not as much on servicing existing customers. Consider this example. When you call your phone company, cable provider, or such commoditized product providers you typically hear two main options. If you are a new customer, please press 1 and if you are an existing customer, please press 2. Have you ever noticed that pressing 1 almost always gets your call answered much more quickly as opposed to pressing 2? Why does that happen? Isn't it obvious?

Each person in a solution chain has a critical role to play in a company's and its customers' operations. It takes many people with various skill sets to forge ahead. Each team member should consider everyone else with the same degree of importance and realize that each member makes unique contributions irrespective of their titles. This culture of equal respect and fairness has to start from those team members that have conventionally *big titles*. Team members with non-C-level titles, usually, look up to their management teams and emulate their behavior. This top down trickling of courtesy and respect should be leveraged in the right way for the right outcomes. Team members with C-level titles have huge responsibilities riding on their shoulders but many do not realize the type and extent of impact they have on others. Many believe that it is their right to have the biggest offices, shiniest gadgets, and the biggest paychecks but most do not realize that they owe their perks and their glory partly to those with non-C-level titles and that a vast majority of the company toils to make that company profitable and that that vast majority should be celebrated as often as possible without treating them like worker bees. Without every person's valuable contribution no company can be profitable.

Technology leaders should imbue their organizations with a culture of courtesy and equality of importance to ensure that they do not respond only to emails and phone calls from team members with big titles. Leaders should set shining examples that involuntarily compel team members to follow the leader and increase camaraderie and productivity in the entire company.

15 NEW AGE IT

WHEN IS THE LAST time you or your IT department as a whole produced something genuinely new? As much progress as has been made in the world of technology does it seem like big corporations seem to be going round and round in circles? They seem to grow bigger but not necessarily smarter or friendlier. On the other hand, many young companies have dramatically changed how we live, as consumers, and such companies usually do not have the money that big corporations do. As a result, they are forced to innovate, do things differently, and more cost effectively. As they say, necessity is the mother of all inventions and this is so apt for newer, younger companies where their survival is based on their ability to innovate and improvise. They have to have a mentality that results in game changing products and disruptive services. On the other hand, bigger corporations that have a steady stream of revenue and people warming their seats for years, if not decades, do not feel the need to innovate because revenues roll in no matter what. And what happens when the revenue stream is interrupted or slows? The classic management response—reduce costs by reducing head count, rather "right sizing the workforce."

Companies have to realize that the best time to innovate, engineer better solutions, and, thus, increase profitability is when time is on your side. The best time to reduce costs, streamline processes is when revenue is rolling in and there is little or no pressure to control costs. Those are the times when people are likely to think clearly and diligently to thoroughly vet out pros and cons. When there's a time crunch guess what happens. People have knee jerk reactions. Poor, incomplete, half baked proposals are created and bad business decisions are made. When times are good companies spend so much time patting each other on the back and adding more and more customers that they forget all businesses are cyclical in nature. Instead of patting each other on the back managers have to remind themselves to temper the celebration. Companies are known to have frenzied reactions during the up cycle as well as the down cycle. Everyone should work hard to be successful, profitable and then enjoy and celebrate their individual and collective successes. However, enjoying the fruits of one's labor and celebrating success should be done in moderation. No one cares if a company has a few highly profitable quarters. Investors, shareholders, management and employees should focus on steady, sustainable growth via consistent execution and planning that takes into account cyclical nature of business. Being cautiously optimistic and having a healthy dose of caution or even pessimism serve well in the long run. Questioning what might be lurking around the corner that could result in significant challenges and disruption should be considered by the Board and C-level managers. Think of Walkman and iPod, Blackberry and iPhone, Myspace and Facebook, USPS, Fed Ex, UPS and Email, gas guzzling SUVs and smaller, fuel efficient cars and the list can go on and on and on. The common theme in all these was that the perceived market leader was caught flat footed when a new market leader emerged with a better, efficient, cost effective product and the consumers simply moved in droves to the new leader leaving the previous market leader completely unprepared and on the ropes. On the other hand, consider moves by News Corp, and Comcast. News Corp bought The Wall Street Journal (WSJ) and Comcast bought a majority stake in NBCUniversal (NBCU). Why were these moves smart, in my opinion? News Corp and Comcast need new and original content for their highly profitable distribution channels. Relying on sources that they had little control over meant they had little control over their own corporate plans and thus their destiny. As a result, with acquisitions of WSJ and NBCU, respectively, News Corp and Comcast have a much greater, if not total, control over their product value chain. In my opinion, these were very smart and strategic moves that were executed with exquisite timing. However, this is not an innovative strategy. Any business degree or any experienced business operations manager will teach you that gaining control over your sources of *raw material* allows you to control timing and cost, which means a greater control over your profitability.

So, how does all this relate to IT? In this context it could apply to IT departments within companies or companies that provide commodity services to other businesses. IT needs to think about its value proposition and plan 3-5 years ahead of time with more than one backup plan. Technologies have evolved and will continue to evolve at breakneck speed but consumer and customer needs are a little   predictable—need to do things faster, cheaper, and better—with improved security. The need to do many things at once, and need to reach everything and everyone from anywhere at any time are, currently, paramount in consumers' minds. Technology provides a toolkit to solve such challenges and capture opportunities but technology does not necessarily provide a readymade solution for all situations. Technology provides super strong bricks but someone still has to lay the bricks together to build the house. This is where smart leadership, management, and professionals come into play. Without realizing what will be needed and how to solve for that need IT might as well surrender. IT has to know what it is good at (other than taking orders) and know what it should or should not do. IT should provide products and services that can be molded into any reasonable shape without sacrificing security and taking on unnecessary risk. Otherwise, anyone who can provide a more cost effective replacement should replace IT. A classic example of this lack of forward thinking can be found in the world of generating business reports and dashboards. Armies of people are dedicated to generating reports and the world of generating reports has gotten smarter in terms of generating more reports faster, accessible remotely and on mobile platforms but they only tell you what happened the previous day or previous month. It's old news. Reports have to get smarter and tell you not only what happened but how to leverage that information for what is likely to happen the next day or next month. Without such intelligence and value addition what real use do such reports provide? If IT can only deliver stale news then the lowest cost provider should be allowed to build such reports and companies can at least save money if they cannot get true intelligence. I am by no means saying that IT has to invent a magic wand to predict the future but IT has to pool its resources with the relevant TAOs and BAOs to figure out what value such reporting functions can provide and how to make it easier for companies to focus on execution by truly providing rich information and not waste time in unnecessary interpretation of data. In this context, the goal should be to generate reports that do more and tell you more and not to generate more reports in less time.

In conclusion, IT has to not only get smarter at doing what it's used to doing but continuously identify ways of providing more predictive information and tell its customers what they would love to know or to have. When expectations are so low that business customers don't even say what they need, technology leaders are doing something terribly wrong. When leaders say aligning IT Strategy with Business Strategy is their number one priority those leaders have already lost their battle and are not fit to lead their respective technology organizations. IT has to challenge itself, be its biggest critic and that critique has to be objective without falling prey to conventional thinking, age-old policies, and reliance on decades-old infrastructure with a focus on turf protection. IT has to learn to deliver a Wow factor in everything it does and allow customers to challenge everyone—leaders, managers, and staff! After all, IT is not an entity unto its own. IT is comprised of leaders, managers, staff and everyone in between. Every person in IT has to improve their image from the inside out. Without individual efforts the overall image of IT will never change. Ask not only what IT should do for its customers but confidently state what Technology can do for its customers.

ABOUT THE AUTHOR

KEDAR SATHE WAS BORN and brought up in India and has a Bachelors in Chemical Engineering from Osmania University College of Technology in Hyderabad, India. He has a Masters in Chemical Engineering from University of Southwestern Louisiana (now, University of Louisiana at Lafayette) in Lafayette, LA (USA). His Master's thesis included laboratory experiments, field research and software development that gave him a unique perspective on delivering quality while keeping an eye on budgets and timelines.

Mr. Sathe has been programming from the age of fourteen when, in the ninth grade, he laid his fingers on a computer–a Commodore 64. Mr. Nagaraj, his late Computer Science teacher in high school inspired and encouraged his interest in the world of software. Mr. Sathe's diverse experience in Engineering, Oil and Gas, and Financial Services give him a unique perspective on impact of technology on business culture and vice versa.

Throughout his academic and professional career Mr. Sathe has developed software to capitalize on business opportunities and solve real life problems. Nothing motivates him more than solving problems and achieving advancements through gains in productivity and efficiency.